A SCIENCE OF GOAL FORMULATION

A SCIENCE OF GOAL FORMULATION

American and Soviet Discussions of Cybernetics and Systems Theory

Edited by

STUART A. UMPLEBY

George Washington University
Washington, DC

VADIM N. SADOVSKY

Institute for Systems Studies
Academy of Sciences of the USSR, Moscow

HEMISPHERE PUBLISHING CORPORATION

A member of the Taylor & Francis Group

New York Washington Philadelphia London

This publication is a result of a project of the American Council of Learned Societies—Soviet Academy of Sciences Committee on the Humanities and Social Sciences, administered in the U.S. by the International Research and Exchanges Board.

A SCIENCE OF GOAL FORMULATION: American and Soviet Discussions of Cybernetics and Systems Theory

1 2 3 4 5 6 7 8 9 0 BRBR 9 8 7 6 5 4 3 2 1 0

Cover design by Debra Eubanks Riffe.
A CIP catalog record for this book is available from the British Library.

Library of Congress Cataloging-in-Publication Data

A Science of goal formulation : American and Soviet discussions of cybernetics and systems theory / edited by Stuart A. Umpleby, Vadim N. Sadovsky.
p. cm.
Includes bibliographical references and index.

1. Cybernetics—Methodology. 2. System theory—Management.
3. Cybernetics—United States—International cooperation.
4. Cybernetics—Soviet Union—International cooperation.
I. Umpleby, Stuart A. II. Sadovskiĭ, V. N. (Vadim Nikolaevich)
Q315.5.S35 1991 90-46513
003—dc20 CIP
ISBN 0-89116-735-8

CONTENTS

PREFACE

This book consists of the proceedings of a meeting among eight American and eight Soviet scientists on the foundations of cybernetics and general systems theory. The meeting was held at Airlie House in Warrenton, Virginia, near Washington, DC, May 17-19, 1985.

The reader might be interested to know how this meeting came about. Our first conversation about a joint project took place at a conference of the Society for General Systems Research held in Toronto in 1981. The intention was to understand better the development and evolution of cybernetics and systems theory in the two countries. Further discussions were held during a conference in West Germany in 1982, in Moscow in 1983, and in Vienna in 1984. We considered a number of methods for exchanging views and finally settled upon a series of meetings involving a small group of scholars from the two countries.

The 1985 meeting was a project under the auspices of the American Council of Learned Societies (ACLS)—Soviet Academy of Sciences Committee on the Humanities and Social Sciences. These projects are administered in the U.S. by the International Research and Exchanges Board (IREX). The Soviet organization which is immediately responsible for this series of meetings is the Institute for Systems Studies in Moscow. In the United States the project is being arranged by members of the American Society for Cybernetics.

The papers were grouped under three topics—epistemology, methodology, and management. The American side was particularly interested in epistemology. Recent interest in cybernetics in the US has focused on observing systems, or constructivist epistemologies, rather than the earlier interest in observed systems, which assumed an objectivist epistemology. The Soviet side was interested in methodology, which they interpreted to include mathematical modeling. Mathematical modeling has long been a central concern for Soviet scientists because of their efforts to improve the efficiency of the central planning process in their economy. Both sides were interested in the management of social systems, particularly group process methods.

As a result of these meetings we are beginning to understand each other better. We have found some common points of agreement and some differences in how scholars in the two countries approach cybernetics, systems theory, and management. Although there is considerable agreement on who the key contributors to the field have been, there is some disagreement on what the important issues are today. This is particularly true regarding philosophical subjects. In addition there seem to be some differences in the complexity of the assumptions made about cognition, with the Soviets beginning from a more complex set of assumptions than the Americans. There is more agreement but not complete agreement on applied topics.

The United States and the Soviet Union are examples of two quite different social systems. Since cybernetics is usually defined as the science of communication and control, these two countries can be regarded as case studies of communication and control in large social systems. More than we realized when we began, we have found that scientific papers need to be understood as expressions of the societies in which they were generated. In addition to the scientific content of these papers, the careful reader may find that the topics discussed and how they are discussed are also informative.

The title of this volume may seem somewhat grand when compared with the content of the papers. There was little explicit discussion of goal formulation during the meeting. However, the participants did repeatedly return to the topic of paradox. At first there was some confusion about what people had in mind when they talked about paradox. As it turned out some participants felt that pointing out paradoxes was a useful way to cause decision-makers to rethink their positions. Other participants were interested in paradox for theoretical reasons. Paradox has been discussed frequently by members of the American Society for Cybernetics in recent years because self-referential statements lead to paradox. Examples of self-reference are a social scientist describing the social system of which he is a part and an artificial intelligence device constructing a description of itself. Paradox is a type of inconsistency, and formal theories are ideally free of inconsistencies. Hence, self-reference appears to impede our ability to develop formal theories of cognition, artificial intelligence, or management. However, the conundrum was laid to rest, at least temporarily, when Ernst von Glasersfeld noted that paradox can be resolved by formulating goals. Without clearly defined goals one gets lost in logical difficulties.

From the earliest writings the field of cybernetics and systems theory has been concerned with issues of governance and purposeful behavior. Although much of the work in this field in recent decades has had a rather narrow, technical focus, broader and deeper questions have motivated the major advances. One American participant suggested at the end of the meeting that the focus of attention within cybernetics and general systems theory has shifted from the analytical techniques associated with achieving goals to the conceptual, philosophical, and ethical problems involved in formulating goals. These two foci of attention were evident at this meeting with the newer concerns being expressed more frequently. Much of the discussion reflected a desire to move from a consideration of techniques to notions of purpose and to approach this subject in a scientific fashion.

Always present in the background was an awareness of the arms race and the ideological competition between the United States and the Soviet Union. These issues were sometimes discussed overtly. At other times they influenced the course of the discussions in more subtle ways.

This volume contains the revised versions of the papers which were presented at the meeting and most of the discussion that occurred during the meeting. The discussion was recorded in writing by the participants during the conference. Hence the discussion presented here does not include everything that was said but does contain those points which the participants were moved to write down. No significant topics discussed during the meeting were eliminated prior to publication.

Of course not all of the trends in cybernetics, systems theory and management were covered at the meeting. Certainly not all of the problems were discussed as deeply as many would desire. But the meeting was a first step. A second meeting is planned for the Soviet Union in 1988. At that meeting we shall discuss four topics—philosophy, methodology, management, and large-scale social experiments. The topic of large-scale social experiments has been added because of the great interest, in the United States as well as the Soviet Union, in the reforms now taking place in the USSR under the name "perestroika."

The Appendix lists the names and current affiliations of the American and Soviet scientists who took part in the meeting.

Stuart A. Umpleby
Washington, Summer 1987

Vadim N. Sadovsky
Moscow, Summer 1987

ACKNOWLEDGMENTS

Many people contributed to making possible this book and the meeting on which it is based. We would like to thank the participants for their papers and their spirited participation in the discussions. The dedicated people at the International Research and Exchanges Board and the Soviet Academy of Sciences provided the necessary funds and handled travel arrangements. Numerous people at George Washington University worked long hours making preparations for the meeting, particularly Fred Eberhard and Sandra Pornovets.

Three of the papers included in this volume are edited versions of papers published elsewhere. For permission to reprint the papers of George Klir, Ian Mitroff, and Milan Zeleny, we thank the publishers of *Systems Research*, *Technological Forecasting and Social Change*, and *The International Journal of General Systems*.

Florence Padgett at Hemisphere Publishing Co. was unfailingly cheerful and helpful during the preparation of the manuscript.

Group's photograph. (*Back row*) George Klir, Milan Zeleny, Dmitry A. Pospelov, Igor A. Apokin, Michael U. Ben-Eli, Ernst Von Glasersfeld, Vadim N. Sadovsky, Vladimir S. Rapoport, Oleg A. Kossov, Vladimir M. Munipov, Nicholai I. Lapin, and Bruce Allyn. (*Front row*) Igor A. Lavrov, Ian I. Mitroff, William J. Reckmeyer, and Stuart A. Umpleby.

EPISTEMOLOGY

CHAPTER

1

KNOWING IN SELF-REGULATING ORGANISMS: A CONSTRUCTIVIST APPROACH

Ernst von Glasersfeld

The earliest historical records we have of questions concerning knowledge and the activity of knowing are from the 6th century B.C. The thinkers who raised them are now usually referred to as pre-Socratics because, at least in the modern world, Socrates was famous before them.

What has been preserved from the pre-Socratics is nothing but a few fragments, the testimony of other writers, and the not always friendly references made to them by Plato in his dialogues. Nevertheless, there seems to be quite enough to show that several among these versatile men had no illusions about the possibility of "objective" knowledge. They realized full well that, once the knowing subject was separated from the objects that were to be known, there was an insurmountable logical obstacle. If the subject's "true" knowledge was to be a picture of the real world, one could never check its "truth" because there was no way of comparing the subject's knowledge with the objects it was supposed to depict.

In the history of Western theories of knowledge, the sceptics have never tired of repeating that argument and the professional philosophers have twisted and turned to find a way around it. They have come up with several very beautiful and very inspiring attempts, but it would be difficult to maintain that they have solved the problem.

The startling thing about this is that, although we cannot prove the truth of our knowledge, we seem to have a remarkably stable experiential reality in which we carry on with our daily living. We formulate explanations, we make predictions, and we even manage to control certain events in the field of our experience. Not only that. We also have scientific knowledge, and it seems to be the most solid. We rely on it, and it allows us to do many quite marvelous things.

For epistemologists, then, it may be useful to look at how we come to have that kind of knowledge. From my point of view, Humberto Maturana has provided the most lucid analysis of the procedure that is usually called "the scientific method."[1] Maturana divides the procedure into four steps:

1. OBSERVATION. In order to count as "scientific," an observation must be carried out under certain constraints, and the constraints must be made explicit (so that the observation can be repeated).
2. Observations may then be related by an HYPOTHESIS, usually an inductive hypothesis that involves causal connections.
3. By deduction a PREDICTION is derived from the hypothesis, a prediction that concerns an event that has not yet been observed.
4. The scientist then sets out to observe the predicted event; again, the OBSERVATION must take place under certain explicit constraints.

Throughout the four steps, what matters is experience. Observing is a way of experiencing and, to be scientific, it must be regulated by certain constraints. The hypotheses by means of which one then relates one's observations, connect experiences, not "things-in-themselves." The predictions, again, regard experiences, not events in some "real" world that lies beyond one's actual experience.

Seen in this way, the scientific method does not refer to, nor does it need, an ontological reality—it concerns exclusively the experiential world of observers.

Scientific knowledge, then, does not and could not yield a picture of the "real" world; it provides more or less reliable ways of dealing with experience. Hence it may be viable, but it can make no claim to "Truth," if "Truth" is to be understood as a correspondence to the ontologically real world. On the other hand, this way of looking at knowledge, be it scientific or other, makes it immune against the sceptics' perennial argument. Since this constructivist notion of knowledge does not claim to provide a picture of something beyond experience, the fact that one cannot compare it with such a something, does not detract from this kind of knowledge—it is either viable or it is not. Indeed, as a constructivist, I tend to go one step further: Since we have access only to experience and cannot get outside the experiential field, there is no way one could show that one's experiences are the effects of causes that lie outside the experiential world.

Saying that we have no cognitive access to a world beyond our experience may suggest to some that any such theory of knowledge must be very close to, if not identical with, the dreaded heresy of solipsism. I do not think that this applies to the kind of constructivism I have been promoting. The reason is simply that, although we deny the traditionally posited iconic relation between knowledge and ontological reality, we substitute for it a different but no less specific relation. Unlike the notion of "truth," which requires a match, i.e. shared points, between the picture and what it is to represent, the notion of "viability" requires fit which, in this context, could be characterized precisely by the absence of shared points. The concept of "viability," however, does imply that there are or will be obstacles and constraints that impinge on whatever aspires to becoming viable. It is not the case that "anything goes," and it is precisely through its obstructions that ontological "reality" manifests itself: by impeding some of our actions and by thwarting some of our efforts. The salient point in all this is that, since this "reality" manifests itself only in failures of our acting and/or thinking, we have no way of describing it except in terms of actions and thoughts that turned out to be unsuccessful.

However, even if we accept this way of thinking for the moment, we would like to have a kind of knowledge that we can call "objective" as opposed to some lesser kinds of knowledge. I have elsewhere proposed a model that provides "objectivity" at the highest level of experiential reality.[2] Needless to say, this constructivist objectivity does not lie in or even point to a world of things-in-themselves, but lies wholly within the subject's field of experience, because it arises through the corroboration of "Others" which the subjective observer can construct within his or her own experiential domain. This construction is, in fact, an extension of a suggestion Kant made in the 1st edition of his Critique of Pure Reason:

> "If one conceives of another thinking subject one necessarily imputes to that other the properties and capabilities by which one characterizes oneself as subject."[3]

This creation of Others in our likeness does not happen all at once. It begins quite harmlessly with the child imputing the capability of spontaneous movement to items in the experiential field that do not stay put. It is followed by the imputation of visual and auditory senses to animals, and it is crowned by the imputation of goal-directed behavior, deliberate planning, and experiential learning to Others whom one considers "like" oneself. Once this level of sophistication is reached, one spends a great deal of time explaining, predicting, and attempting to control these Others. That is to say, one now has models of moving, perceiving, planning, thinking, feeling, and even philosophizing Others in one's experiential field, models to whom one imputes the kinds of concepts, schemes, and rules one might oneself abstract from one's experience. These models incorporate some of the knowledge we ourselves have found useful and thus viable in our own dealings with experience. If, then, we are able to make a successful prediction about any one of these Others, the particular piece of knowledge which, in making the prediction, we have imputed to the Other, acquires a second order of viability: we now feel justified in saying that this piece of knowledge was found to be viable not only in our own sphere of actions but also in that of the Other. This, I believe, is as close as a constructivist can come to "objectivity."

It is obvious that the construction of a viability of which I can say with some justification that it seems to reach beyond my own field of experience into that of Others, must play an important part in the stabilization and solidification of my experiential reality. In fact, it helps to create that highest level of which we then believe that it is shared by Others and, therefore, "more real" than anything experienced only by ourselves.

This kind of corroboration, one might think, is much more easily and much more generally achieved by linguistic communication. From the constructivist point of view, however, the notion of "sharing" as a result of a linguistic exchange turns out to be the result of the very same kind of imputation we have discussed above. This brief summary is not the place to expand on the constructivist approach to language and communication.[4] All I want to say here is that, in spite of the fact that it often feels as though the meaning of words and sentences were conveyed to us by the sounds of speech or the signs on a printed page, it is easy to show that meanings do not travel through space and must under all circumstances be constructed in the heads of the language users. If we then ask, what these meanings could be made of, we find that the only raw material available is the stock of experiential records the individual language user has so far accumulated. There is no doubt that these subjective meanings get modified, honed, and adapted throughout the course of social interaction. But this adaptation does not and cannot change the fact that the material of which these meanings consist can be taken only from the individual language user's subjective experience. (Note that, in this respect, social adaptation is analogous to biological adaptation: it can do no more than bring out, recombine, or thwart what is already in the organism—it cannot instill new elements.)

It may be useful to repeat that constructivism does not deny reality, nor does it deny that the living organism interacts with an environment; but it does deny that the human knower can come to know reality in the ontological sense. The reason for this denial is simply that the human knower's interactions with the ontic world may reveal to some extent what the human knower can do—the space in which the human knower can move—, but they cannot reveal the nature of the constraints within which the human knower's movements are confined. Constructivism, thus, does not deny the "existence" of Others, it merely holds that insofar as we know these Others, they are models that we ourselves construct.

In the light of what was said above about the levels of the reality that we construct of our experience, Others are of very great importance: without them, we could not attain the highest level of that experiential reality. Thus, we need Others.

In the Western philosophical tradition, attempts to create a rational basis for ethics have never succeeded in finding an unquestionable justification for the initial assumption that we ought to consider Others as we consider ourselves. The model of the cognitive organism as a self-regulating entity that builds up "knowledge" as ways and means to maintain itself in whatever space the ontic world has left for it to live in, this model seems to justify the human organism's need for Others and, in doing so, it justifies the first requirement of any ethics on epistemological rather than on ethical grounds.

This, it seems to me, could have some far-reaching consequences. Once Others are seen as an epistemological requirement needed for the construction of our own experiential reality, we may be able to counteract some of the influence the concept of competition has wielded for so long owing to its unfortunate and largely erroneous association with the theory of evolution. Similarly since the notion of "controlling others" tends to destroy the possibility of corroborating the viability of one's own cognitive structures, the constructivist approach to ethics would foster tolerance.

When competition and control of others are eliminated, we may come to have an interactive pattern of mature humanity. That pattern I would call "collaborative dialectic," because it would be very much like the pattern of biological evolution which, according to the most recent findings, seems far less competitive than co-evolutionary.

Conclusion: If knowledge can be considered the result of the adaptive effort of cognitive organisms in their struggle to maintain their equilibrium in the face of perturbations, it does not seem reasonable for them to use this knowledge to compete with one another. On the contrary, it seems that in order to maintain not only their own equilibrium but also that of the planet on which they find themselves living they would have to foster in every conceivable way every kind of mutual collaboration.

REFERENCES

1. Maturana H. & F. Varela, *The Tree of Knowledge*, Boston: Shambhala, New Science Library, 1987.
2. von Glasersfeld, E. Steps in the construction of "others" and "reality": A study in self-regulation. In *Power, Autonomy, Utopia: New Approaches toward Complex Systems*. New York: Plenum Press, 1986.
3. Kant, I. *Kritik der reinen Vernunft*, (1. Auflage, 1781). Berlin: Akademieausgabe, Bd.IV, p. 223.
4. von Glasersfeld, E. On the concept of interpretation. *Poetics*, 1983, 12, 207-218.

DISCUSSION OF VON GLASERSFELD'S PRESENTATION

Apokin

1) Let us suppose that your viewpoint is the opposite. That is, human experience has ontological substance. What would be the result for your model of cognition? Would the result be the same?

2) The same question, but concerning ethical results.

Von Glasersfeld

Whether human experience has or has not ontological substance could be decided only by a god (an entity that is in a position to compare human experience *and* ontological reality). My point is simply that the belief that the knowledge gained from experience is a "true" reflection of an absolute reality tends to make people dangerous—they begin to think that their knowledge is the only "right" or "true" knowledge. With a constructivist perspective our interpretation changes from "this is the way it is," to "this is one way that works."

Mitroff

If Kant were here, he would probably say things like, "The statement 'observation of...' already implies that one has made an ontological commitment; i.e. 'things' called observations are posited to exist."

Von Glasersfeld

Adopting a constructivist orientation entails, as I suggested earlier, changing some important notions. "Observing" and "observation" are certainly among those that have to be changed. In the realist view, "to observe an object" implies the independent existence of that object. Constructivists are much closer to Kant and, speaking for myself, I would say that "to observe something" means no more than to focus one's attention on a particular area of what Kant called "the manifold" and what present-day neurophysiologists would call "events in the organism's nervous system."

Sadovsky

How do we acquire objective knowledge?

Von Glasersfeld

Objective knowledge is not a metaphysical assumption. It is only a way of dealing. Objective knowledge, for a constructivist, is not a necessary metaphysical assumption. Once one relinquishes the idea that knowledge *must* depict something ontological, and once one defines knowledge as ways of acting and/or thinking that *fit* into the world rather than replicating it, there cannot be any objective knowledge in the iconic or correspondence sense.

Sadovsky

Is there not a contradiction between recognizing, on the one hand, that knowledge is not the reflection of some independently existing reality and, on the other hand, attempting to find ways of constructing objective knowledge (as you do)? Do you perceive difficulties of principle in employing the criterion of community agreement for constructing objective knowledge? The history of science yields a number of negative examples. In your opinion is there a way of resolving the paradox of knowledge, for example, if we take a historical approach to the development of knowledge and consider its practical results? And is a constructive approach really possible in realism? In my view, a union of realism and constructivism is quite possible through dialectics.

Von Glasersfeld

Your question is actually several questions:

1) Is there a contradiction between (a) the conception of "knowledge" as a cognitive construction without iconic relationship to "some independently existing reality" and (b) the construction of "objective" knowledge?

Answer: No, because I define "objectivity" in a way that is different from the tradition.

2) Do I perceive difficulties in employing the "criterion of community agreement" in the construction of "objective knowledge"?

Answer: No, because I do not speak of "community agreement". In my model, "other people" are my own construct, just as are "tables", "chairs" and the "pencil" I write with, the "trees" and "stars" outside. These (and all other items) I construct on the basis of (a) repeated coordination in the sensory mode, (b) correlation of two or more sensory modes. From these constructs and the way I relate them, I abstract "operational" concepts of "space", "time", "motion", "state", "change", "causality", etc., that are the building blocks of "physics" and "physical knowledge". This construction is subject to selection, the sense that I maintain only what "works", that is, what serves any purposes within the constraints of my experiential world. Whatever has proven "viable" in my experience up to now, I would call "knowledge".

There are items in my experiential world that do not strictly obey the rules I have abstracted as "physical rules". Like every child, I discovered that I could make some special rules for these extraordinary items, if I hypothesized that they could see, hear, feel, and think in the way I do. Thus I have "models" of other people that are more or less viable—in the sense that they allow me to predict, explain, and sometimes control them. But the viability of these models is the result of selection (i.e., of other models having failed to "work") in the same sense as in the case of my "physical" models of things and their interactions. But there is a difference: because I have imputed perceiving, thinking and planning to these "others", whenever a prediction about them turns out to be "correct", this confirms my model of them and, therefore, my way of perceiving, thinking and planning. This confirmation allows me to call my models, and the hypotheses they confirm, "objective".

3) What if we take a historical approach to the development of knowledge and its practical results?

Answer: All sorts of things may happen. Above all, the registering of repetition will make the repeatedly satisfactory constructs more "viable". It may lead to "habits" of perceiving and thinking. But it does not bring my constructs closer to ontological reality because it does not make them less my constructs; and the notion of "history" is, after all, also my construction (e.g., "time", "change", etc.).

4) Is the constructive approach possible in realism?

Answer: If "realism" implies the belief that our knowledge does or could iconically reflect ontological reality, no.

But that does not mean that for a constructivist there is no ontological reality. I merely say that the relation between it and my knowledge is not iconic or a relation of correspondence. My *experiential* reality is constructed within the constraint of the ontological world, and I cannot *know* these constraints because they manifest themselves only in the *failure* of my actions.

5) If "realism" were defined as assuming the reality of the constraints within which we act," and if we call "real" the hypothetical model we may construct out of the failures of our actions, and if we define "dialectics" as the interaction of our present knowledge with that model of "reality", then constructivism and realism can be "united".

Lapin

1) The idea of coevolution instead of competition is very interesting. But coevolution can be based on various types of interdependence: subordination or cooperation among elements of the system. Do you accept this?

2) Hypothesis testing (4th stage of the scientific method) can be immediately objective, for example, a lunar flight confirms the flight computations. Hence this experience at once becomes objective. Do you agree?

3) The first stage (observation) assumes that in addition to the observer there is the observed object, i.e. something objective (other than cases of self-observation). Do you accept this ontological premise?

Von Glasersfeld

1) Interdependence, yes, in the sense that both co-evolved organisms (species, societies) have developed a way of surviving that requires the survival of the other.

2) No, because the lunar flight is an *experience* and no experiential confirmation of an hypothesis can provide evidence of an ontological reality that would have to be posited *beyond* the domain of experience.

3) No, in my (and Kant's) way of speaking "experience" is subjective and *it would be unwarranted to posit objective causes of experience.*

I try not to talk about existence but rather about knowledge. Constructivism is different from other philosophical perspectives. The emphasis is on "fit" or finding ways of managing, not on "being".

CHAPTER

2

PHILOSOPHICAL AND METHODOLOGICAL FOUNDATIONS OF SYSTEMS THEORY

Vadim N. Sadovsky

1 INTRODUCTION

Today, 35-40 years after the appearance of the first works on general systems theory and cybernetics written by L. von Bertalanffy and N. Wiener who pioneered in these fields of knowledge, nobody casts doubt on the fact that the systems movement is an essential component of modern science, technology and different forms of practical activity. The systems movement can be rightfully regarded as a manifestation of the contemporary scientific and technological revolution. Practically all scientific and technological disciplines are drawn into it. It in equal measure touches upon both research and development. Under its influence the latest conceptions of planning and management are developing and methods for solving global problems are being elaborated.

Being interdisciplinary by its nature, contemporary systems research has a complex hierarchical structure, including both ultimately abstract, purely theoretical and philosophical-methodological components, and numerous practical applications. In this paper the object of analysis will be only one side of systems research, namely, the complex of philosophical problems that have arisen in the process of formation and development of systems research methods. These problems really exist and they are rather important. This idea was recognized from the first moments of the systems movement, but, unfortunately, for a long time appropriate attention has not been paid to this aspect of systems research. The situation changed only at the close of the 60s and at the beginning of the 70s, when in different countries in the context of different philosophical traditions numerous programs for investigating the philosophical problems of systems research were suggested. Rather active development of this sphere of systems research began, but we can hardly be satisfied with the results of this process which are available now.

If we try to give the most general appreciation of the present-day level of investigation of the philosophical problems of systems research, we can characterize it as follows: 1) as a period of advancement of alternative development programs for this field which, as a rule, are weakly connected with each other; 2) as a period of lack of terminological rigor and clarity, when every author usually defines not only the whole complex of main systems concepts, but also many of the most important philosophical concepts and principles all over again; in a word, an initial, even preparatory, if you like, period of scientific development of systems research. But I think it is time to try to insert the necessary rigor and systemicity into this field.

2 SHORT HISTORICAL EXCURSUS

As is well known, the original works on general systems theory (Bertalanffy, 1945; 1950) and cybernetics (Wiener, 1948) were of purely scientific character. Though each of these theories pretended to be of a very high (if not universal) level of generalization, their creators were very careful to attach deep philosophical significance to them. So, Bertalanffy, who in the 40s and 50s was inclined to regard general systems theory as a step toward creating a universal science embracing all other sciences, later, in particular in response to our (together with Lektorsky, 1960) criticism concerning this thesis, stressed that he was not going to assert that general systems theory played the role of "the philosophy of contemporary science". To his mind, general systems theory in its present form is one—and still a very imperfect—model among other analogical ones; it will never be "exhaustive, exclusive or final" (Bertalanffy, 1962). There was an analogous situation in the early stage of the development of cybernetics.

In this connection it is interesting to mention that A. A. Bogdanov who at the beginning of the 20th century created tektology—a universal organizational science which without any doubt became the historical predecessor of both general systems theory and cybernetics—regarded tektology as a science, not a philosophical conception (Bogdanov, 1980). This affirmation is sufficiently forceful, because it was expressed by a professional philosopher and, moreover, it represents to a considerable extent a conclusion of the complex evolution of Bogdanov's philosophical views.

But in spite of the above mentioned view expressed in the 40s and 50s by Bertalanffy and Wiener regarding the attachment of philosophical significance to general systems theory and cybernetics, the first steps to reveal the philosophical meaning of these new fields of scientific and technological knowledge were made in the works on general systems theory of A. Rapoport (1950, 1963), K. Boulding (1956), J. Miller (1955) and others published within the same period and in the works on cybernetics of Wiener (1948, 1955), W. Ross Ashby (1958), H. von Foerster (1960) and others dated from the 50s and 60s.

A similar process was taking place in this period in Soviet investigations of cybernetics and general systems theory. A. A. Lyapunov and A. I. Kitov (1961) gave the original account of the scientific and theoretical contents of cybernetic principles and models of organization, control and communication. In the first Soviet works on general systems theory (Kremyansky, 1958; Lektorsky, Sadovsky, 1960) the main attention was paid to laying bare the generalized scientific essence of this conception. At the same time in these works a complex of new philosophical problems was defined, problems which arose in connection with the rise of such general scientific theories as general systems theory and cybernetics and whose solution was necessary for progress in these fields. It should be noted that rather soon, i.e. in the middle of the 60s already, in the USSR philosophical problems of cybernetics had been formed into quite an autonomous field of philosophy with large numbers of publications, with courses of study and so on (see Berg and others, 1962). At the same time assessing in retrospect the degree of comprehension and the level of investigation of philosophical problems of systems research which were reached in the 60s, we must regard this period as just the first stage of development of the philosophical sphere in systems research.

A new stage in the elaboration of this complex of problems began at the close of the 60s and at the beginning of the 70s. It differed from the preceding period of the recognition of the philosophical significance of systems research, because concrete programs of investigation in this field of knowledge were formulated. We should use the term "programs", not "program", because investigations which had similar tasks were carried out practically simultaneously in different countries in accordance with different philosophical traditions. All this led to the rise of essential differences in these programs.

In the English-speaking countries the program for creating systems philosophy proposed by E. Laszlo (1972) became widely known. This program caused very soon a wide reaction, both positive and negative (see, for example, the journal "Metaphilosophy", 1972; Berlinski, 1976). It was this program which influenced L. von Bertalanffy, and he changed his view of the philosophical meaning of general systems theory. In his last works, some of which were published after his death, L. von Bertalanffy, defining the main direction of the development of systems research, distinguished side by side with systems science and systems engineering, systems philosophy which he defined, following E. Laszlo, as consisting of three relatively autonomous parts: systems ontology, systems epistemology and a systems theory of values (L. von Bertalanffy's preface to E. Laszlo's book, 1972).

At the close of the 60s within the framework of Marxist philosophical concepts, a program for developing the philosophical problems of systems research had been formulated. By the 70s a great many scientists were taking an active part. V. G.

Afanasiev (1973), J. M. Gvishiani (1979), I. V. Blauberg, V. N. Sadovsky, E. G. Yudin (1969, 1973), V. P. Kuzmin (1976) and many others had convincingly shown that the adequate philosophical foundation of systems research was materialist dialectics and that the clue to understanding the whole complex of philosophical problems of systems research was a deep comprehension of the meaning of the philosophical systems principles developed by K. Marx. Interesting results in this field were also derived by A. I. Uemov (1978) who built the so-called parameterical variant of general systems theory and investigated the philosophical meaning of relational generalizations, G. P. Shchedrovitsky (1974), who analyzed the methodological nature of the systems movement, and some other Marxist philosophers. I. V. Blauberg, V. N. Sadovsky and B. G. Yudin in 1978 (English translation, 1979) showed the difference between the systems approach as a general scientific and methodological conception of systems research (not pretending to solve the philosophical problems proper) and the philosophical systems principle as the central principle of a specifically philosophical analysis of systems research methods and the systems view of objects of the real world.

In the second half of the 70s further progress in the development of systems philosophy took place. P. M'Pherson (1974) in his assessment of perspectives of contemporary science and systems philosophy gave a generalized account of E. Laszlo's approach. In a program of creation of general systems theory as philosophy T. Downing Bowler (1979) proposed another method for substantiating systems philosophy, connected mainly with an attachment of great methodological importance to relational generalizations. Simultaneously there appeared programs of development for systems philosophy proposed by M. Bunge (1977, 1979), A. Bahm (1970, 1981) and some others.

The present-day state of systems philosophy or philosophy of systems is rather adequately reflected in an analytical review of "The Systems Movement" (published by Cavallo, 1979): practically nobody has any doubt of the importance of this sphere of systems research, but everyone elaborating it goes along his own path being worried neither about connections between his conception and other similar ones, nor about progress in this field of knowledge as a whole. Moreover, achievement of mutual understanding is hampered by terminological discord, lack of rigor in the use of key concepts and so on. Using systems language, we can convincingly affirm that a non-systems approach dominates in the present-day investigation of the philosophical problems of systems research. Of course this state of affairs cannot be regarded as normal. The first task to be solved, in this connection, concerns discussion of the status and substantiation of the program for developing systems philosophy.

3 CRITICAL ANALYSIS OF THE PROGRAM FOR CONSTRUCTING "SYSTEMS PHILOSOPHY"

In the research recently conducted by Marxist philosophers they convincingly demonstrated that the philosophic foundation of systems studies is materialistic dialects, in particular, a philosophic systems principle that represents a facet or aspect of the dialectic method (Kuzmin, 1976; Gvishiani, 1984; Sadovsky, 1984b). In addition, many western scholars have tried to solve the problem of philosophic substantiation of systems by constructing a special "systems philosophy". This is, in their opinion, a contemporary form of philosophic knowledge that overcomes the limitations of the earlier philosophic conceptions and is consistent with the scientific and technical achievements of the second half of the 20th century. I would not like to cast any doubt on the fact that in each historical period of its existence philosophy should integrate the achievements of science and technology by application of its methods and conceptual structures (the results of systems studies, in particular, are expressed and substantiated in a philosophic systems principle). I believe, however, that the construction of a special "systems philosophy" is not promising. This opinion may be reinforced by the following arguments:

a) The numerous versions of "systems philosophy" mentioned earlier (E. Laszlo, T. Downing Bowler, M. Bunge, A. Bahm, et al.), different as they may be, have a point in common: their unjustified assumption that systems philosophy is a special scientific knowledge.
b) In the ontological interpretation, "systems philosophy" in the final analysis reproduces the patterns of constructing *Naturphilosophie* whose scientific and heuristic value is known to be insignificant.
c) In the epistemological interpretation, "systems philosophy" does not contain any distinctive features that would make it possible to treat this stage of philosophic evolution as a qualitatively new form of philosophy—systems philosophy (see for more details Sadovsky, 1985).

Therefore further progress in understanding the philosophic foundations and meaning of systems studies should not, in my opinion, be sought in the construction of a special "systems philosophy". In order to achieve progress, one should specify the content and meaning of the philosophic systems principle as a part of dialectics.

4 ELEMENTARISM, HOLISM AND THE PHILOSOPHIC SYSTEMS PRINCIPLE

In the dialectic materialistic substantiation of systems studies one takes into account not only the "whole-parts" dichotomy but also the source of a system's change or evolution and methods of understanding.

There are two radically different views on problem solving: recognizing the primacy of 1) the parts over the whole; or 2) the whole over the parts. As to the source of a system's change or evolution, it may be understood as a) an integral ideal factor; b) a self-initiated movement that is possible due to the material nature of the systems under review; c) an external (with respect to the given event) material force.

Finally, there are two major approaches to the selection of preferable methods of research: one of them emphasizes the importance of (I) analysis, the second of (II) synthesis.

The above-mentioned basic solutions of philosophic problems connected with the dichotomy of the whole and parts, with the identification of the source of a system's change or evolution and methods of understanding form three fundamental philosophic approaches. The first of them—let us call it *elementarist*—recognizes the primacy of elements (parts) over the whole, identifies the source of a system's evolution in the activities of external objects (with respect to the object under consideration) and regards analysis as the only method of cognizing the world. Historically the elementarist approach appeared in various forms, and each of them specified in this or that way the common features of elementarism.

The second fundamental philosophic approach—it would be right to call it *holistic*—is based on the recognition of the primacy of the whole over the parts. It regards the source of a system's evolution as being some integral and, as a rule, ideal factors. It prefers synthesis as a method of studying objects. There is a great variety of "shades" of holism—from sheer idealistic vitalism, J. Smuts holism that hardly differs from it, to quite respectable (scientifically) conceptions of emergentism and organicism.

The third fundamental philosophic approach is the *philosophic systems principle*. It recognizes the primacy of the whole over the parts but simultaneously emphasizes the interrelationship of the whole and the parts expressed, in particular, in the hierarchic structure of the world. The source of a system's change or evolution is considered to be self-initiated movement. The essential condition of adequate knowledge is the unity of analysis and synthesis in the strictly rationalistic interpretation (not intuitive). The essence of the systems principle may be reduced to the following features: 1) the integral nature of the external world's objects and subjects of cognition; 2) the interrelationship between the elements of any object and the given object, on the one hand, and numerous other objects, on the other; 3) the dynamic nature of any object; 4) the functioning and evolution of any object as a result of its interaction with the environment, primacy belonging to the internal regularities of the object (its self-initiated movement) over the external ones.

The systems principle understood in this way overcomes the limitations of elementarist and holist conceptions in all their numerous versions (Rapoport, 1986) and therefore turns out to be the most adequate philosophic foundation of systems studies.

In addition to the solution of these problems, the philosophic systems principle makes it possible to determine the methodological structure of the systems approach that we would like to present as follows:

$$S = \langle W, M, P, R, \alpha, Str(Org), Hier, E, G, B, I, C \rangle$$

This scheme expresses the system's essential variables and the requirements of the systems approach. In a most general way they may be presented as: the identification of integral properties (*W*) of the system under review and its numerous disaggregations into elements (*M*); traditional study of the properties (*P*), relationships (*R*) and connections (α) of the given system with other systems, as well as its subsystems, parts and elements, identification of its structure/organization *Str(Org)* and its hierarchical structure (*Hier*). The last group of methodological requirements of the systems approach and, consequently, the system's properties is concerned with the determination of the relationships between the system and its environment (*E*), the goals of the system and its subsystems (*G*), description of the system's behavior (*B*), including its development, identification of the informational aspect of the system (*I*) and the control of the system based on the information circulating in the system and its environment (*C*). It is essential to point out that all the above methodological requirements imply a unity of analysis and synthesis as methods of research. They also express the unity of internal and external stimuli in the system's functioning and development.

Thus, the suggested methodological structure of the systems approach expresses the essential components of the philosophic systems principle. This scheme is proposed as a useful methodological guide in specific systems studies.

REFERENCES

1. Afanasiev, V. G., On the Systems Approach in the Social Sciences, *Voprosy filosofii*, N6, pp. 98-111, 1973 (in Russian).
2. Afanasiev, V. G., *Systemicity and Society*, Politizdat Publishing House, Moscow, 1980 (in Russian).
3. Ashby, W. Ross, General Systems Theory as a New Discipline, *General Systems*, vol. III, pp. 1-6, 1958.
4. Bahm, A. J., *Polarity, Dialectic, and Organicity*, Charles C. Thomas, Springfield, Illinois, 1970.
5. Bahm, A. J., Five Types of Systems Philosophy, *International Journal of General Systems*, vol. 6, N4, pp. 233-237, 1981.
6. Berg, A. I., et al., Cybernetics, in *Philosophical Encyclopedia*, vol. II, pp. 495-506, Moscow, 1962 (in Russian).
7. Berlinski, D., *On Systems Analysis: An Essay on the Limitations of Some Mathematical Methods in the Social, Political, and Biological Sciences*, MIT Press, Cambridge, Massachusetts, 1976.
8. Bertalanffy, L. von, Zu einer allgemeinen Systemlehre, *Blätter für deutsche Philosophie*, N 3/4, 1945.
9. Bertalanffy, L. von, An Outline of General System Theory, *British Journal for the Philosophy of Science*, vol. 1, pp. 134-164, 1950.
10. Bertalanffy, L. von, General System Theory—A Critical Review, *General Systems*, vol. VII, pp. 1-20, 1962.
11. Bertalanffy, L. von, *General System Theory: Foundations, Development, Applications*, Allen Lane, The Penguin Press, London, 1971.
12. Blauberg, I. V., Sadovsky, V. N., Yudin, E. G., *The Systemic Approach: Prerequisites, Problems and Difficulties*, Moscow, 1969 (in Russian); English Translation—*General Systems*, vol. XXV, pp. 1-31, 1980.
13. Blauberg, I. V., Sadovsky, V. N., Yudin, E. G., Some Problems of General Systems Development, in *Unity through Diversity*, A Festtschrift for Ludwig von Bertalanffy, ed. by W. Gray and N. D. Rizzo, Part I, pp. 245-270, Gordon and Breach, New York, 1973.
14. Blauberg, I. V., Sadovsky, V. N., Yudin, B. G., Philosophical Principle of Systemicity and the Systems Approach, *Soviet Studies in Philosophy*, vol. XVII, N 4, pp. 44-68, 1979.
15. Blauberg, I. V., Mirsky, E. M., Sadovsky, V. N., Systems Approach and Systems Analysis, in *Rethinking the Process of Operational Research and Systems Analysis*, ed. by R. Tomlinson and I. Kiss, pp. 19-32, Pergamon, Oxford, 1984.
16. Bogdanov, A. A., *Essays in Tektology: The Universal Organization Science*, ed. by G. Gorelik, The Systems Inquiry Series, Intersystems Publications, 1980.
17. Boulding, K., General Systems Theory—the Skeleton of Science, *General Systems*, vol. I, pp. 11-17, 1956.
18. Bowler, Downing T., General Systems Theory as Philosophy, in *Improving the Human Condition: Quality and Stability in Social Systems*, Proceedings of the Silver Anniversary International Meeting of the Society for General Systems Research, London, August 20-24, 1979, pp. 20-29.
19. Bunge, M., General Systems and Holism, *General Systems*, vol. XXII, pp. 87-90, 1977a.
20. Bunge, M., The GST Challenge to the Classical Philosophies of Science, *International Journal of General Systems*, vol. 4, N 1, pp. 29-37, 1977b.

21. Bunge, M., *Treatise on Basic Philosophy*, vol. 4, *Ontology II: A World of Systems*, D. Reidel, Dordrecht, 1979.

22. Cavallo, R. E. (ed.), Systems Research Movement: Characteristics, Accomplishments, and Current Developments, *General Systems Bulletin*, Special Issue, vol. IX, N 3, 1979.

23. Foerster, H. von, On Self-Organizing Systems and their Environments, in *Self-Organizing Systems*, ed. by M. C. Yovits and S. Cameron, Pergamon Press, Oxford, 1960.

24. Gvishiani, J. M., Materialist Dialectics as a Philosophical Basis for Systems Research, in *Systems Research, Methodological Problems*, ed. by J. M. Gvishiani, Pergamon, Oxford, 1984.

25. Kremyansky, V. I., Certain Peculiarities of Organisms as a "System" from the Point of View of Physics, Cybernetics, and Biology, *General Systems*, vol. V, pp. 221-230, 1960.

26. Kuzmin, V. P. *The Systems Principle in K. Marx's Theory and Methodology*, Politizdat Publishers, Moscow, 1976 (in Russian).

27. Laszlo, E., *Introduction to Systems Philosophy, Toward a New Paradigm of Contemporary Thought*, Gordon and Breach, New York, 1972.

28. Lektorsky, V. A., Sadovsky, V. N., On the Principles of Systems Research (Related to L. von Bertalanffy's "General System Theory"), *General Systems*, vol. V, pp. 171-179, 1960.

29. Lyapunov, A. A., Kitov, A. I., Cybernetics in Technics and Economics, *Voprosy filosofii*, N 9, pp. 79-89, 1961.

30. Metaphilosophy, Papers by E. Laszlo, L. von Bertalanffy, L. Thayer, S. Pepper, and R. Burhoe, *Metaphilosophy*, vol. 3, pp. 123-155, 1972.

31. Miller, J., Towards a General Theory for the Behavioral Sciences, *American Psychology*, vol. 10, pp. 513-531, 1955.

32. M'Pherson, P. K., A Perspective on Systems Science and Systems Philosophy, *Futures*, June, pp. 219-239, 1974.

33. Ogurtsov, A. P., On Understanding of the Systems Nature of Scientific Knowledge (Antiquity and the New Times), in *Systems Research, Yearbook 1974*, Nauka Publishers, Moscow, pp. 154-186, 1974 (in Russian).

34. Rapoport, A., Horvath, W. J., Thoughts on Organization Theory and a Review of Two Conferences, *General Systems*, vol. 4, pp. 87-94, 1959.

35. Rapoport, A., Remarks on General Systems Theory, *General Systems*, vol. VIII, pp. 123-124, 1963.

36. Rapoport, A., *General System Theory: Essential Concepts and Applications*, Abacus Press, Cambridge, MA, 1986.

37. Sadovsky, V. N., Systems Approach and General Systems Theory—Status, Main Problems, and Prospects for Development, in *Systems Research, Methodological Problems*, Pergamon, Oxford, 1984a.

38. Sadovsky, V. N., Problems of the Philosophical Foundation of Systems Research, in *Systems Research, Methodological Problems, Yearbook 1984*, Nauka Publishers, Moscow, pp. 32-51, 1984b (in Russian).

39. Shchedrovitsky, G. P., *The Systems Movement and the Future of Systems Structure Methodology*, Obninsk, 1974 (in Russian).

40. Uemov, A. I., *The Systems Approach and General Systems Theory*, Mysl Publishers, Moscow, 1978 (in Russian).

41. Wiener, N., *Cybernetics or Control and Communication in the Animal and the Machine*, MIT Press, 1948.

42. Wiener, N., *The Human Use of Human Beings*, New York: Avon-Hearst, 1950.

DISCUSSION OF SADOVSKY'S PRESENTATION

Reckmeyer

A dumb question: Would you clarify your definitions of, or patterns which connect, the following fundamental terms?

1. Philosophy
2. Epistemology
3. Ontology
4. Methodology

Sadovsky

A wise answer: Would you be so kind as to consult an encyclopedia?

A little more serious answer:

1) Philosophy's main tasks are the establishment of general laws of nature, society and the development of the human mind. Philosophy is the basis of the scientific world outlook. In philosophy we deal with subject—object relationships as well as with epistemological, ethical and aesthetical relations.

2 and 3) Ontology and epistemology are traditional and essential parts of philosophy. The former is concerned with the explanation of Being, the latter with understanding knowledge activity and the results of knowledge. It is very important that in both cases we deal with the ultimate conditions and limitations of Being, Human Existence and Knowledge.

4) Methodology is one part of epistemology. Its main aim is the construction of rules and procedures for different kinds of investigations—theoretical, practical, technological, etc. Methodology as well as epistemology itself has an ontological basis.

Allyn

What is the difference between two cases which you have mentioned in your presentation: 1) the systems approach is a part of dialectics, and 2) dialectics is a philosophical basis for the systems approach?

Sadovsky

The difference is quite clear. In the first case dialectics and the systems approach are considered to be at the same level of methodological thinking. From my point of view this is not adequate. In the second case dialectics is considered to be a philosophical methodology and the systems approach belongs to a different level of methodology—to the level of general scientific methodology.

CHAPTER

3

THE COMPLETE AND UTTER FAILURE OF TRADITIONAL THINKING IN COMPREHENDING THE NUCLEAR PREDICAMENT: WHY IT IS IMPOSSIBLE TO FORMULATE A PARADOX-FREE THEORY OF NUCLEAR POLICY

Ian I. Mitroff

The radically different nature of nuclear weapons, the radically different type of world of which they are a partial reflection, and that they have helped to usher in, reveal that ordinary processes of thinking are completely inadequate to assess the nature of these weapons and to formulate strategies appropriate for dealing with them. The reason is that the dialectical contradictions and paradoxes inherent in the very nature of nuclear weapons and the doctrines concerning their use are far deeper and vastly more extensive than has been previously recognized. To be sure, the paradoxical nature of nuclear weapons and theories of deterrence have been recognized from their very

Adapted and reprinted by permission from *Technological Forecasting and Social Change*, 29, 51-72 (1986).

inception. I am thus not contending that the recognition of paradox and internal contradictions per se is anything new. What I am contending is that because of the dramatically different nature of today's world, a world that nuclear weapons have played a significant role in ushering in and in shaping, it now appears that it is not possible to formulate *any* theory or policy of nuclear strategy that will be free of significant, extensive, deep paradoxes and contradictions. The search for perfectly consistent, contradiction-free, and paradox-free theories or policies for controlling or limiting the use of nuclear weapons seems doomed to utter hopelessness. This does not mean that there is a lack of all hope and that there are no theories which are better than others on some grounds yet to be specified. What it does mean is that no theory or policy however preferable it might be will ever be free of significant paradoxes and internal contradictions. Therefore, if an evaluation and a final choice is to be made between different policies, it will have to be made on grounds vastly different from what has been argued by analysts thus far [1].

There are four principal categories into which the paradoxes and contradictions associated with nuclear thought fall. It is both interesting and important to note that the categories not only capture the specific and detailed paradoxes and contradictions that operate *within* each of the separate categories but that the system of categories taken as a whole captures the paradoxes and contradictions that operate *between* the categories considered as generic types. In addition to the four main types, there are also some miscellaneous categories. When one is dealing with a phenomenon of such complexity and magnitude, it would be the height of folly to suppose that one could force fit it for the sake of some compulsive notion of order down to four neat categories. In the same spirit, none of the categories pretends to strict exclusiveness and, least of all, complete exhaustiveness. Phenomena like nuclear war are the premier example of what have been termed, to put it mildly, "ill-structured" or "wicked" problems, i.e., phenomena that unlike textbook exercises do not admit of a single definitive formulation, let alone a single, exact solution [2]. The essence of the phenomena cannot, in short, be fully captured by any single schema. Unlike well-structured problems whose definition and solution are presumably the same for all experts, the definition and solution of ill-structured problems are variable depending on the particular interested party. For this reason alone, they cannot be resolved by such simple epistemic decision mechanisms as "widespread consensus." For if their resolution were dependent on such mechanisms, then taking for example the single issue of abortion, we should have learned by now that before we even begin our investigation, widespread consensus will never emerge. Therefore, we need different decision mechanisms that are more epistemically appropriate for shedding light on issues that are truly wicked, i.e., that are racked by deep divisions of opinion and conflict. Once again, we need different ways of thinking about such issues if we are even to grasp the "true" nature of their complexity. The following categories are thus offered in the spirit of shedding important light on the nature of this complexity, not in the spirit of providing the definitive last word on it.

The four principal categories are: (1) more is or leads to less; (2) less is or leads to more; (3) more is or leads to more; and (4) less is or leads to less.

1 MORE IS LESS

Table 1 is a listing of *some* of the many key issues that are involved in various aspects of nuclear policy. It makes no pretense to completeness whatsoever. Its main purpose is to give an indication of the variety of paradoxes that are involved in some of the most important issues connected with nuclear policy. Indeed, the items in all the tables, not merely those in Table 1, were selected to illustrate as many diverse facets of nuclear policy as possible. Also, in nearly every case, each item is a direct or near direct quote from the huge literature on nuclear policy. I have thus deliberately adopted the tactic of using quotes to illustrate that the existence of paradoxes and contradictions is not a fabrication of my making. Indeed, as profuse as the number of items in the tables are, they are but a bare pittance by comparison to all of the paradoxes and contradictions that could have been listed. In fact, as we shall see later, *there is Not A Single Proposition that can be uttered with regard to nuclear policy that does not have a significant paradoxical quality to it or for which there is not another proposition which is the extreme dialectical opposite of it.*

Table 1

MORE LEADS TO LESS

Theme	Proposition
Superiority	1. a. Decisive superiority leads to greater vulnerability. b. The more nearly equal two sides are, the greater their mutual vulnerability [3].
Superiority	2. Clear superiority (e.g., even in the '60s when US had it) does not necessarily lead to greater knowledge regarding how to exploit it [4].
Flexibility	3. Flexible policy responses might lead other side to keep one step ahead in the escalation ladder and thus go higher in its responses than it intended [5].
Flexibility	4. Greater flexibility of nuclear options leads to the greater illusion of the control of nuclear strategy [6].
Verification	5. When increased satellite reconnaissance became coupled with increased missile accuracy, greater verification of weapons became as destabilizing as it became stabilizing; i.e., it leads to greater vulnerability [7].
Accuracy	6. Inaccurate plus hi-yield missiles indicate a strategy of deterrence (i.e., less & more leads to more); high accuracy missiles indicate a war fighting strategy, and therefore are more provocative (i.e., more leads to less) [8].

Defense	7. Fortress hardening of missile silos against surprise attack only promotes the arms race; it challenges the enemy in precisely the area where he's ready to respond [9].
Global	8. Better US/USSR arms control treaties could free more USSR weapons to go to the Far East [10].
Bargaining	9. Negotiating from strength may lead to weapons programs (bargaining chips) that no one wanted in the first place but which consequently develop a strong constituency; no administration can tell taxpayers it's spending a fortune on chips whose chief purpose is to be given away; weapons are selected as chips when the point of no return is passed in their nonemployment [11].
C3I Vulnerability	10. The more sophisticated the command, communication, control, information (C3I) system for controlling attacks, the more vulnerable it is to attacks [12].
Safety	11. Greater safety measures increase the weight of missiles and jeopardize missiles' workability [13].
Beliefs About Winnability	12. US beliefs that nuclear war is winnable may make war more likely which may make Soviets reason similarly which may make war more or less likely [14].
Fragile technology	13. Our greatest strength is our strong technological base which is more vulnerable than a preagrarian society to which we can't go back; we have created a more fragile infrastructure [15].
Management of Secrecy	14. High secrecy leads to making worst case assumptions about the enemy which fuels the arms race [16].
Management of Threat	15. The greater the threat one delivers, the greater the loss of one's subsequent choices of policy options [17].
Proliferation	16. a. Greater proliferation of arms is to be encouraged for it concentrates people's minds wonderfully (more leads to more) b. versus its unwarranted gamble; because it has worked, no guarantee it will continue (more may lead to less) [18].
Rationality	17. Rationality may be neither necessary nor sufficient for deterrence; irrationality (less) could lead to passive acquiesce (less); while rationality (more) could lead to belligerence (less) [19].

History	18. More experience (history) with nuclear weapons has not lead to any more sensible options than were available 30 years ago [20].
ABM futility	19. It is futile to spend billions on antiballistic missile defense because Soviets could counter with increased missile offense at less cost [21].
Trade	20. Increased US/USSR trade made US farmers more dependent on USSR; therefore, it immobilized trade as a weapon on USSR [22].
Trade	21. Greater control of products that USSR wants from US improves USSR efforts to be independent, brings USSR defects to their attention; hence, we are their monitoring system [23].
Oversell of USSR threat	22. Must oversell USSR threat to get big defense $ in a democracy; but counterproductive since simple arguments ultimately draw attention to dilemma of extended deterrence (i.e., our commitment to defend Europe and not merely the US against a nuclear attack) which leads to backpeddling [24].
Linked weaknesses	23. Democracies must publicize weaknesses to get more defense $ which undercuts our deterrence; USSR exaggerates its strengths which fuels our hawks [25].
Weapons	24. Weapons aimed at weapons (counterforce) increase risk of war; weapons aimed at people lessen risk of war, therefore are better [26].
Management of Terror	25. The perennial dilemma: no one knows how to plan a strike big enough to be terrifying but small enough to be recognized unambiguously as limited and hence not prompt a big counter-strike in return [27].
Management of Risk	26. a. Deliberately raising risk of all-out nuclear war actually lessens the possibility of war; thus, decreases vulnerability. b. But this leaves everyone less sure that war can be kept under control; thus, greater risk leads to greater vulnerability [28].
MIRVs	27. a. MIRVed missiles are more cost effective; therefore better off (more leads to more). b. worse off because a more valuable target. c. worse off because not as accurate as a single missile. d. better off because one missile can attack several targets [29].

Arms Control	28. Greater quantitative limits have accelerated the qualitative arms race [30].
Arms Control	29. Present arms control agreements don't halt the arms race but institutionalize and regulate it so it can continue with less instability [31].
Arms Control	30. Arms control talks require both sides to be sensitive to military comparisons that are often irrelevant [32].
Arms Control	31. Stressing importance of arms control talks may lead to greater disillusionment when they fail to produce [33].
Arms Control	32. Differences between USA/USSR cultures are so deep that arms control processes may serve to increase suspicion, not lessen it [34].
Joint Chiefs	33. SALT III if it occurred could bankrupt the US because of sweetener payoffs to joint chiefs for going along with treaty [35].
Better Earthquake Prediction	34. Better methods for predicting or causing earthquakes could dramatically increase the number of violations of underground nuclear test explosions prohibited by treaties. The reason is that one could mask the shock waves set off by a prohibited explosion by those produced by a naturally occurring hazard [36].
Proliferation	35. Deterrence makes the proliferation of nuclear weapons more likely the more that deterrence is seen to succeed [37].

With very few exceptions, all of the items in Table 1 satisfy the following pattern: what initially seems to be a good, desirable characteristic or property of a nuclear policy leads upon further reflection or deeper examination to a paradoxically undesirable or bad outcome; in other words, something initially perceived to be good leads to something bad or undesirable. The generic expression that best seems to capture the nature of this paradoxical twist is: MORE LEADS TO LESS.

The items or issues per se listed in Table 1 necessitate little comment. They are intimately familiar to those who are conversant with nuclear policy. For example, item 1a is one of the classic and first-recognized dilemmas in nuclear policy, i.e., the dominant superiority of one side tempts the other less superior side to strike first. Further, since each side's arsenal is presumed to be lethal, the concept of superiority loses much of the meaning it once had in earlier times. At the same time, item 1b asserts that being approximately equal is also just as dangerous because in this century wars have broken out between two sides just as much under a condition of near equality as when one side was far ahead of the other. Thus, no matter which way we turn, we are caught in a dilemma: (1) being ahead, (2) being equal, and (3) being behind all are bad, if not equally so. What's going on here? What has changed about the nature of the world to make this so?

All of the paradoxes in Table 1 arise from the radically different nature of nuclear weapons acting in a radically different kind of world. The sheer number of nuclear weapons and the sheer magnitude of their destructiveness makes them truly global weapons; they are truly weapons in the systemic sense. They act on the entire planet considered as a whole system. At the same time, this has occurred precisely when the entire planet has been moving towards an underlying infrastructure that is itself highly interconnected. Today there is no part of the infrastructure of modern societies that is not tied in to all of its other parts: transportation, communications, education, energy, agriculture, entertainment, etc. Each affects the others in ways we only partially understand. Further, as the Bhopal and Chernobyl tragedies show the infrastructures of all societies are more connected than ever before. What happens in a plant in a so-called underdeveloped part of the world affects the headquarters of a multinational company located in a so-called developed society. In a world that is more and more tied together and affected by what goes on in every part of the globe, the terms "underdeveloped" and "developed" lose much of their former meaning. The U.S. is highly developed in some aspects and woefully underdeveloped in other aspects. One of the major consequences of this is that *all* of the distinctions that were once relevant in a simpler world are now absolutely meaningless. They totally collapse. For instance, Andrei Gromyko was absolutely correct when he argued that space-based anti-missile defense is an "insidious ruse." "If the United States had such a shield against enemy missiles, [Gromyko] said, it could fire its own missiles without fear of retaliation." [38] The distinction between offense and defense that once was relevant is now without foundation.

Another consequence is that things that were once highly desirable in a simpler world where countries were more disconnected from one another by the fact of geographical distance (e.g., the numerical superiority of men and arms in the conventional sense) are now highly undesirable in a world that is increasingly one big interconnected system. But if this is so, then the failure to trace through characteristic properties that seem desirable "in the small" and in isolation can be disastrous. It is only in a world conceived in the image of a simple machine that effects can be partitioned off from one another and studied in isolation from one another. Only in this kind of world can problems likewise be separated off and solved in isolation from one another. This is because it is implicitly assumed that they can be removed from one another like the separate parts of a machine and studied in isolation. Today, increasingly none of the problems of the world either exist or can be studied in isolation from all other problems. A mechanistic (i.e., reductionistic) view of problems not only fails to treat them properly but actually makes them worse.

For instance, a simple but dramatic example is item number 34 in Table 1. On the surface of it everyone would agree that better prediction of earthquakes would be a good thing. And in a world without global weapons it would be. But if we or the Russians could predict earthquakes more accurately, then either of us could use this information to time unauthorized underground nuclear tests such that the shock waves produced by the test explosions would be masked by the shock waves produced by an earthquake of the right magnitude. Jervis has put the whole matter well:

> Superiority without the ability to protect one's civilization does not give either side much leverage. Drawing even with the Soviets—assuming that they are indeed ahead of us— would not alter the vulnerability of American cities, and it is this which inhibits the United States from using force under circumstances which could lead to all-out war [39].
>
> Asking who would come out ahead in a nuclear war, either in terms of residual military and political power, assumes a zero-sum situation. Such an assumption made sense in eras when the costs of waging war were not prohibitive. The side that was ahead not only was better off than its adversary, but was less likely to be better off than it would have been without a war. Now however, even the side that is "ahead" of the other after a major war will be much worse off than it would have been had peace and the status quo been maintained. Again, the prenuclear ways of thinking have been carried over into a context in which they can only mislead us [40].

In sum, things in the past that counted for superiority (greater numbers of men, arms, territory won and occupied) have lost much, if not all, of their former meaning. The things that worked and counted in the past only did so because the side with greater superiority could use it to inflict devastating damage on the other side without suffering from the rebound effects such devastation caused its enemies. In short, the initiator was insulated through distance from the devastation it wrought on its opponents. Today, whether through the vulnerability of a lethal second strike or the possibility of a nuclear winter, neither side is immune from the other, or more strongly, from itself. In a world where one is likely to suffer as much from the devastation one can inflict on one's enemies, one has as much to fear from oneself as from one's enemies.

In a world that is increasingly fashioned in the image of a complex system, judging whether something is therefore a desirable feature or not in the nuclear era cannot be rendered in isolation from the consideration of other features, features that by themselves were desirable in the prenuclear era. It can only be judged in terms of its consequences and implications for the planet considered as a highly complex, whole system.

Notice carefully that this does not mean that even in the nuclear era every one of the seemingly desirable characteristics listed in Table 1 are totally bad or completely undesirable per se in all contexts. For instance, no one would abandon or deny the desirability of arms control talks per se, or for another, the need for mutual verification of some kind. My point is that above all in the nuclear era the seemingly desirable properties of the prenuclear era cannot be judged or evaluated in isolation from the total package (i.e., policy) of system effects in which they are embedded. Also they cannot be considered as properties that are so inherently desirable in themselves that any policy that is lacking in them is immediately suspect or without virtue. As so many have said, thinking about war in the prenuclear era does not transfer to the nuclear era. The type of thinking that is appropriate for the nuclear era is thus precisely what this article is about.

Another way to put the matter is to say that none of the seemingly desirable characteristics in Table 1 can be maximized without limit or careful consideration of their qualitative effects on the whole. The preeminent social systems thinker Russell L. Ackoff may have put it best of all in his carefully drawn distinction between "growth" and "development" as often opposing criteria in terms of which to measure the performance of a system [41]. Too many institutions strive for the unrestricted growth of some measure, typically quantifiable, as the sole legitimate measure of performance. Thus, for instance, if the rate of return on investment of a corporation was 15% last year, then it should strive for 20% in the next. Ackoff contrasts this with the concept of development which is the pursuit of goals or ends which are typically nonquantifiable. Development typically has more to do with an individual's or a system's overall increase in its "felt or perceived quality of life" as opposed to the mere increase in some numerical indicator, e.g., unemployment. In this sense, development often has more to do with the heightened awareness of an individual's aesthetic sense in his or her entire life. Ackoff puts it well when he bitingly points out that cemeteries grow each year but they don't develop. The same is true of trash heaps. The questions that must be reserved for later after we have explored the other categories of paradoxes is: what would a nuclear policy look like that was oriented more toward the concept of development, and not merely that of growth?

2 "LESS IS MORE"

Table 2 lists some of the various aspects of nuclear strategy that can be associated with the generic heading LESS LEADS TO MORE. Again, there is no presumption of complete coverage or exclusiveness of the items.

There are a number of points in Table 2 that are worth commenting on. One, nearly every one of the items is highly contentious, as are all the times in all the tables. I do not for one moment expect that everyone would necessarily agree with my classification of them. But then the basic point of the tables is not to secure widespread agreement, since this is generally not possible in ill-structured problems. Instead, their function is to force recognition and provoke discussion of an important paradoxical aspect of nuclear strategy. That others would conceivably place different items in their tables is desirable if this in itself would help to make possible a different level and kind of discussion than has taken place thus far.

Table 2

LESS LEADS TO MORE

Theme	Proposition
Uncertainty	1. Uncertainty (less certainty) is precisely what has kept the nuclear peace for 30 years; both sides have a stake in keeping the other from believing it really knows how

		weapons would perform; don't know conclusively how weapons would behave until fired but unless have conclusive knowledge no rational leader is likely to risk firing [42].
Control, Uncertainty	2.	Does a trip wire outside of our control lessen or increase the changes of war (i.e., does less control lead to more or to less?) [43]?
Limits, Knowledge and Rationality	3.	a. We know how to start but not how to end a nuclear war; i.e., little knowledge leads to less.
		b. Rationalist thinkers have difficulty in understanding 3a. since they believe that cost benefit analyses of damages by protagonists will stop further irrationality, i.e., less will lead to more.
		c. But limits are more confusing during war; limited war is no longer limited; i.e., less leads to less [44].
Weakness Can Be Strength	4.	Not always an advantage to be rational; may be perfectly rational to wish oneself not altogether rational; in bargaining weakness is often strength, e.g., burning one's bridges behind oneself to demonstrate utter commitment to one's enemy [45].
Mutual Understanding	5.	a. Concepts of deterrence have to be mutually understood and accepted to work; must be a generally accepted assessment of adversary's most valued assets that both parties will accept (thus, more leads to more).
Absurdity and Credibility		b. A nuclear threat which is absurd and unreal becomes more credible and real (i.e., less rationality leads to more).
Cool Judgment-Fright (Oxymoron)		c. Deterrence works from frightening the adversary but requires cool judgment even in a crisis (i.e., less and more lead to more) [46].
Discretion Threat	6.	For maximum credibility, essential to leave as little room for discretion in carrying out the threat [47].
Aggression and Fear	7.	Defensive motivated aggression is more frequent and important than cold-blooded power aggression in starting conflicts; therefore, we should encourage realistic but discourage exaggerated fear (more and less lead to more) [48].
Unilateral Negotiation	8.	Unilateral negotiation by one side may provide just the coordination that will save both parties [49].
Crisis Management	9.	In an era of nuclear weapons, some crises and crises management have become a substitute for war [50].

Tension Reduction	10. In nuclear weapons, a great deal less than equality may be sufficient for deterrence; approximate equality without tension reduction is dangerous [51].
Minimum Deterrence	11. Case for minimum deterrence: in vast oceans, 5 nuclear subs hard to find and could destroy 200 largest Soviet cities [52].
Weapon Nondifferentiation	12. In order to encourage weapon reduction, treat all weapons as alike; let each side turn what it considers most vulnerable [53].
Who's "Ahead"?	13. Who's ahead in arms race is whoever strikes first, not just numbers. (i.e., less leads to more, but does it when it is a "more" which leads to "less"?) [54].
Counterforce and Stability	14. a. If advantages of a first strike can be eliminated or reduced, then incentive to strike at all reduced (less leads to more); thus, weapons of great counterforce power upset stability of balance of terror (more leads to less). b. But if both sides are essentially "invulnerable" to a "lethal" (contradiction in terms?) first strike, if counterforce is destabilizing, if massive retaliation is suicidal, then how could US use threat of nuclear attack to deter Soviets in Europe? (more leads to less) [55].
Slow Missiles	15. Tomahawks are highly mobile, have a low flight path, can sneak in under air defenses with pinpoint accuracy but slow speed, thus not a first-strike weapon [56].
Advantages of Simple Warheads	16. MIRVed ICBMs doubly destabilizing; simultaneously threaten and invite a preemptive attack (more leads to less); thus, single warheads are preferable; too numerous and widely dispersed to preempt; also, smaller, mobile, therefore harder to knock out [57].
Control of a Limited War	17. a. No possibility of controlling a limited war (i.e., less leads to less) versus b. chances of terminating a limited war better than a few years ago (i.e., less leads to more) [58].
Conventional Arms	18. A conventional buildup is desired over a nuclear one since it raises the nuclear threshold [59].
USSR View of US ICBM Vulnerability	19. USSR views growing vulnerability of US ICBMs as stabilizing rather than dangerous [60].
Prestige	20. Some states have obtained prestige by not developing nuclear weapons; if the superpowers halted the arms race,

		this would create a moral climate that would undermine the prestige of having nuclear weapons [61].
Technical Backwardness	21.	Some technical backwardness is an advantage; early USSR ICBMs were large to deliver heavy warheads; as systems improved, heavy throw weight became useful in allowing high-yield warheads and MIRVs [62].
Arms Talks & Disillusionment	22.	USSR has not shown the disillusionment over arms control because it has not expected as much as the US [63].
Treaty Violation	23.	Can't be sure a treaty is worth giving up without violating it [64].
Treaties	24.	Treaties with limited duration invite Soviets to break out the moment they expire (less leads to less) versus they give flexibility (less to more) [65].

Two, there is no way of avoiding basic issues of semantics. An appreciation of this is itself one of the major purposes of Table 2. The issue of semantics concerns the fact that the words "less" and "more" are used in multiple senses within the table. For instance, in some cases the term "less" refers to the fact that less of something initially thought to be desirable, like certainty in item 1, is felt by some analysts to lead to something desirable or "more" in the end. In other cases, like item 9, "less" refers to a carefully managed amount of something perceived to be undesirable at first. I have therefore not striven for semantic consistency because I do not believe this is possible. Indeed, it is the search for simple criteria such as consistency that has hampered our development of policies for containing our nuclear predicament.

Three, and this is one of the most important features, I have deliberately included in the table some items which explicitly go against the grain of the "less leads to more" pattern. I have deliberately done this because it is not possible in many cases to gain a full appreciation of what "less leads to more" means except by contrasting it with its opposites. It is really impossible to keep the contradictions and opposing opinions which rack the theories of nuclear doctrine strictly apart in separate tables.

Four, and this is the most important point of all, in a number of cases I have deliberately placed opposing points of view side by side not only to illustrate the nature of the deep dialectical opposition which haunts all theories of nuclear doctrine but to illustrate something much more profound. Consider items 1 through 5. It is not merely that there are deep divisions of opinion over say whether certainty or uncertainty (or rationality versus irrationality) leads to better control of nuclear weapons, i.e., deterrence. Rather, the case is that in many situations we know no more today than we did thirty years ago when many of the theories of nuclear deterrence were first proposed, which of two opposing pairs of propositions regarding human behavior will apply or result in a given situation. For instance, consider item number 4. No one would deny that it is not always an advantage to be rational. Be this as it may, this is somewhat beside the point if it does not miss a fundamental point about human behavior altogether. Many, if not most, of the theories of nuclear strategy we have are the products of those with an extreme rationalist bent of mind. That is, the theories we have are the products of those who think of human behavior in chess-like terms, i.e., if we do such and such, then the enemy will *rationally* calculate the costs and benefits to him of his various responses and choose the one that is most appropriate. The basic trouble with this is not that people are not rational. It is rather that they are complex mixtures of both rational and irrational impulses which act simultaneously. But to appreciate this one has to have at the very least a minimum understanding of social thinkers such as Freud. In contradistinction to those disciplines which emphasized only the rational side of man's nature, Freud and others emphasized that man was continually buffeted by opposing instincts, an internal dialectic as it were, that influenced all his internal and external affairs. To think that nuclear doctrine would be free of these influences when the opposite is so patently the case is thus the height of folly.

But this is precisely why current theories of nuclear strategy or deterrence are so frightening. We just don't know-period!—whether for example a state of uncertainty will frighten someone into rationality or into its opposite of irrationality. The whole enterprise rests on too many shaky, unknown, and dubious assumptions to be taken with any degree of confidence. We don't even know whether our lack of confidence regarding our lack of confidence, etc., is itself good or bad. It is not therefore that we lack all knowledge whatsoever. Item number seven in Table 2 is about as good and as established a proposition in the social sciences and which pertains directly to the field of nuclear strategy as we are likely to have. It is just that too much else remains frustratingly unknown on which to base any truly reliable theory of nuclear strategy.

3 "MORE IS MORE" "LESS IS LESS"

Table 3 and 4 affirm all of the points made with regard to Table 2. One, all of the items are again contentious. Two, there are issues of semantics once again. While "more" of something generally means something better or a more desirable state, it depends as everything else from whose perspective will generally be clear so that something better from the perspective of the Soviets will not necessarily be better from ours although it can be.

Three, items are included in each table which go against the main thrust or grain of the table to illustrate once again the intense differences of opinion which exist on nearly every, if not virtually all of the issues. And four, there is as before a deep dialectical stand-off between competing propositions about human nature and political systems and how both would respond in severe crises.

In addition however, Tables 3 and 4 introduce something which sheds even deeper light on the nature of the phenomenon with which we are dealing. *With the final appearance of all four tables, one is able to begin to see that there is an aspect of every phenomenon connected with nuclear policy that fits in all four tables or categories simultaneously.* This has the most profound consequences for the treatment and understanding of nuclear policy.

Table 3

MORE LEADS TO MORE

Theme	Proposition
Superiority	1. If the USSR eliminated 90% of US's ICBMs, US couldn't use remaining 10% to retaliate because USSR would still be superior; thus, US would simply surrender [66].
Massive Retaliation Dysfunctional	2. Threat of massive retaliation encourages USSR to engage in greater piecemeal aggression because massive retaliation to small scale aggression lacks credibility; side with greater conventional forces can go on making incursions; more leads to more versus more leads to less [67].
Criticality of Assumptions	3. Even modest shifts in assumptions of missile accuracy are sufficient to shift advantages from attacker to defender (i.e., less leads to less versus more leads to more) [68].
Deterrence Works	4. Deterrence has a track record; has prevented nuclear and conventional war as well; one is morally obligated to come up with a better alternative; can go on working provided we stay clearheaded [69].
Drastic Tension Reduction plus Minimum Deterrence	5. To prevent nuclear war, both sides need to combine minimum deterrence and drastic tension reduction (i.e., less and more leads to more), which includes a highly invulnerable 2nd strike capability plus adequate conventional forces; unless tension reduction is drastic by one side, not likely to be noted by other and taken seriously with result both give up [70].
Worst Fears Warrant Big Money	6. To warrant spending money on offense, the dogma in the Air Force was to brainstorm what horrifying things Soviets might be able to do to us and then find evidence to support worst-case fears (more leads to more and/or less leads to less [71].
Soviet Worst Case Fears	7. US didn't try to talk Soviets out of their fears about Pershing II because if US going to give it up better to let Soviets indulge in their worst case fears and hence pay a good price for its cancellation [72].
Arms Control Nonsensical	8. a. Arms control always inherently controversial: nonsensical for a nation to compromise its defense (i.e., more leads to less).
Arms Control Sensible	b. On other hand, senseless not to control arms that could destroy world (i.e., more leads to more) [73].

Joint USSR/US Crisis Center	9. A world that is more interlinked and whose hi-tech is more vulnerable suggests the establishment of a joint US/USSR manned crisis control center [74].
Conventional Arms Control	10. Nuclear arms control should not limit conventional weapons; i.e., more should not lead to less in all cases [75].
Arms Control	11. Arms control means managing the arms race not ending it; arms control defends military status whereas disarmament threatens it [76].
Space Wars	12. Space is ideal arena for show of forces; does not violate national boundaries; doesn't kill people; visible shows at low cost [77].
Advantages/ Disadvantages of Greater Throw-weight	13. Greater accuracy of new missiles make throw-weight of older Soviet missiles obsolete (i.e., more leads to less) but bigger Soviet rockets means can load more MIRVs onto them (i.e., more leads to more) [78].
USSR Prudence	14. It's absurd to ignore longstanding prudence of USSR with regard to nuclear weapons [79].
Legitimate Participation	15. Soviets felt that victory over Germany earned them the right to participate legitimately in shaping post-war international order [80].
Preparation for War and Deterrence	16. War can be prevented in Soviet thinking only if USSR prepared to fight a war [81].
Proliferation Encouraged	17. a. More nuclear nations reduce probability of war by providing more restraining forces; i.e., more leads to more.
Proliferation Discouraged	b. More nuclear powers leads to greater uncertainty and hence crises are less calculable; may lead to restraint or victim may strike out blindly starting a world war; i.e., more leads to less [82].
Soviet Doves	18. a. Greater access to West gives USSR more realistic view of US fears about Soviet power; optimistic view enhances influence of Soviet Doves to make USSR more reasonable, i.e., more leads to better.
Soviet Hawks	b. Pessimistic: greater understanding strengthens Soviet hard-liners for intervening in Western policies, allows for sharper intrusions rather than blunter ones [83].

Table 4

LESS LEADS TO LESS

Theme	Proposition
Dysfunctionality of Pride	1. We and our opponents are so driven by pride and fear of humiliation that we may threaten force and have to carry it out [84].
Security	2. Security is not a zero-sum game; if we had it but not others, then others would rob us of it; i.e., either more leads to more or less leads to less [88].
Uncertainty Pro and Con	3. Sometimes creating uncertainty in adversary's mind can enhance deterrence but misperceptions may lead to unintended confrontation, failure of deterrence, war (i.e., less can lead to more but it can lead to less) [86].
Lack of Agreement on Basic Data	4. Without agreement on basic data, neither size of necessary reductions or verifications can be agreed on [87].
Worst-Case Thinking As A Self-Fulfilling Prophecy	5. In absence of certainty, safe thing is to overestimate to cope with possibility of greater than expected threat; but, a self-fulfilling prophecy [88].
Inconclusiveness of Who's Ahead	6. The inconclusiveness in the list of pluses and minuses in who's ahead allows for worst case thinking which dominates both sides; such a list now shows that one will never be greater or ahead [89].
Danger of Near Equality	7. In our century, major wars have occurred more often when there was a near approximate equality than when one side was much greater than other; WWI occurred when the balance was more nearly equal than during peace before [90].
Danger of Near Equality	8. The elements of military power both tangible and intangible are so many and varied that any approach to equality creates ambiguity, makes wishful misperception easy [91].
Arms Control Talks	9. To suspend arms control talks when Soviets misbehave suggests that arms control is a favor we do them; plays into hands of hard-liners [92].

USSR/USA Joint Consultation	10. Fact that USSR was not consulted about A-bomb guaranteed that attempts made after WWII to establish international control of nuclear weapons were doomed [93].
Self-Defeating Nature of Limited Nuclear War	11. A limited nuclear war in Europe might destroy what it is supposed to protect [94].
Limited War	12. Concept of limited war makes it more thinkable [95].
Weak Conventional Forces	13. By weak conventional forces, Great Britain makes nuclear war more likely; have to go nuclear weapons sooner; given inherent incredibility of GB nuclear scenarios, is nuclear deterrence a useless and expensive white elephant which weakens rather than strengthens GB forces [96].
First-Use Doctrine	14. Doctrine of no 1st use makes wars more likely; for 30 years war deterred at its origin, prospect even slightest conventional conflict might escalate into a nuclear war so neither has allowed it to happen [97].
First-Use	15. No first use pledge is unenforceable; who will guarantee that loser will stick to it? [98].
Vulnerable 1st Strike Systems	16. Vulnerable weapon systems with a first strike capability might provoke their own 1st use in order to avoid being wiped out; i.e., more and less leads to less [99].
Failure of Deterrence	17. To say not prepared to contemplate failure of deterrence is to make failure more likely [100].
Disarmament	18. It is not a foregone conclusion that disarmament leads to stability; i.e., not clear that less necessarily leads to more; have to be a complete success in order not to be disastrous [101].
Winnability of Nuclear War	19. Even though Soviet view is that nuclear war is madness (i.e., more leads to less), to argue that not winnable is morally dangerous to morale of Soviet troops (i.e., less leads to less) [102].
Lessened Impact of Less Visible Systems	20. Because sea-launched cruise missiles are submerged they would not have desired visibility and political impact of land based systems [103].

One, it means we are dealing with a phenomenon for which ordinary logic just does not apply. The phenomenon just doesn't sit still or fall neatly into one category of thought as it were. It can't be captured by a single system of thought no matter how appealing that system is. That is, unlike ordinary phenomena for which something either is an apple or it is not, for nuclear phenomena something can be both desirable and undesirable at the same time: more can lead to less and simultaneously to more, etc.

Two, it means that there will be no such thing—*ever*—as a policy of nuclear strategy that will only yield positive benefits ("more") without at the same time yielding significant negative consequences ("less").

Three, it means that the kind and level of uncertainty associated with nuclear strategy is much deeper and more pervasive then previously recognized. Not only is there surface uncertainty, or at the very least severe debate, regarding the desirable versus undesirable effects of uncertainty as an important element in policies of deterrence. But at one level down, there is a deeper kind of uncertainty which haunts all theories of nuclear strategy. I know no better term for this kind of uncertainty than that of "meta-uncertainty."

Meta-uncertainty is uncertainty regarding which table(s) or category(ies) ANY specific proposition regarding nuclear strategy will fall into. That is, given the extreme and enumerable uncertainties connected with the large number of propositions or assumptions upon which all theories or nuclear strategy rest, we cannot be sure of which extreme benefits versus extreme costs we will reap once those strategies are put into action.

4 FURTHER PUZZLES AND PARADOXES

If we have demonstrated anything at all, it is that *there is not a single proposition connected with nuclear strategy for which we cannot find a completely opposite proposition that has as much surface credibility or plausibility as the initial proposition.* Further, we have also demonstrated that *there is not a single aspect of nuclear strategy that is free of significant paradoxes and contradictions.* As we have shown these paradoxes and contradictions are of two major kinds: something initially thought or perceived to be good leads to something bad or undesirable, and vice versa.

Table 5 shows that in addition there are some further puzzling, paradoxical, or contradictory properties connected with traditional doctrines of nuclear strategy. One must be both sane and insane, rational and irrational, and *at the very same time*, for deterrence to work, to be credible, etc. To put it mildly, this is either compartmentalized or schizoid thinking of the worst kind. So-called professional, expert analysts may think they are able to handle the contradictions by keeping the various parts of the arguments strictly apart in different unrelated compartments of their minds, but the great psychoanalysts of the human mind taught us that this is fundamentally impossible. *To say that we must be both rational and irrational, sane and insane, at the same time for deterrence to work is really to say that we must be irrational and insane.* There are some contradictions for which one cannot have it both ways.

Table 5

MISCELLANEOUS ASPECTS OF DETERRENCE

Theme	Proposition
Nuclear Scholasticism, Metaphysics	1. More than any other area of public policy, nuclear issues are influenced by theoreticians; because deterrence has never failed, concepts have never achieved status of applied science [104].
Technology Overrides Ideology	2. a. Leveling effect of technology and strategic doctrines overrides differences in politics, social, and ideological systems; actors in MAD essentially unaffected by ideological or value differences [105].
Technology Reflection of Ideology	b. Yet, MAD stands as a partial reflection of US ethos of technology as a decisive shaper of life [106].
Deterrence is Ability to Fight	3. a. Capacity to deter war is logically undistinguishable from ability to fight one [107].
Deterrence Not Equivalent to Use	b. We debate decisions to acquire new weapons as though it is equivalent to use them [108].
Deterrence Not Separate from Use	c. Deterrence operates precisely through capability for actual use; it is not separate from use; weapons that cannot realistically be used cannot deter, lose credibility [109].
Conflict of Functions of Nuclear Weapons	4. Primary function of nuclear weapons is to deter a USSR nuclear attack; 2nd function is extended deterrence of USSR conventional attack on Europe by threatening a US nuclear first use; if nuclear war is apocalyptic, then 1st function contradicts 2nd [110].
Credibility of Deterrence	5. No longer credible that US would invite its own nuclear destruction to stop a Soviet conventional attack on Europe [111].
Credibility and Insane Vindictiveness	6. For deterrence to be credible, not rational political judgments, but insane vindictiveness must be seen as in control [112].

Credibility of of Deterrence

7. Essential aspect of deterrence is that threat be credible to both enemy and ourselves [113].

Intentions to Use Weapons, Not Merely Deter

8. There is a certain logic to the argument if states straining their resources to amass weapons they must be intending to use them, not just deter [114].

Deterrence Not Inherent in Weapons

9. Deterrence is not inherent in the weapons but results from their possession [!] and will to use them [115].

Will to Use What Cannot Be Used

10. Nuclear weapons are useful only to extent never used but adversaries must believe one has the will to use them in retaliation [116].

Morality of MAD

11. Doctrine of Mutually Assured Destruction is moral if it works [117].

Weapons Cause Tensions

12. a. Arms are themselves an independent source of tension, are a root-cause of war because instill mutual fear and hostility [118].

Tensions Cause Weapons

b. Versus assumption of arms control is that weapons themselves don't cause war but are merely symptomatic of deeper tensions [119].

Weapons Remind Us of War

13. Weapons remind us of war because they are parts of our image of war; to think of war is to think of weapons [120].

No Distinction Between Offense and Defense

14. In nuclear age, defense not yet feasible; only possible by deterrence, punishment; defense is offensive capability [121].

No One Knows How to Exploit Advantages

15. In 1961 US had a 10 to 1 ICBM gap in our favor but didn't want even then to strike 1st because too many things could go wrong; didn't know how to exploit advantage then [122].

Endless Cycling Around Various Strategies of Deterrence

16. If whatever US did with nuclear weapons, Soviets could still inflict unacceptable nuclear damage to US, couldn't they use that destructive power as a bargaining tool? Doesn't this force us back to limited war? But limited requires rational military restrictions. Limits are more confusing; therefore limited no longer limited. Therefore are we back to unlimited [123]?

Controlled War

17. It takes two to fight a controlled war; how could we know that the Soviets contrary to everything they have said would play along [124]?

Both Restraining and Driving Forces Needed	18. Chief defect of out-all deterrers (hawks) is that they only focus on negative forces restraining the Soviets, while tension reducers (doves) only focus on positive driving forces; both are needed [125].
USSR Want US Out Europe	19. a. USSR believes that US has no business being in Europe.
USSR Wants US In	b. USSR worries if US were not there that Germans would not be restrained [126].

Now the amount or degree of irrationality or insanity may be up for grabs, but we should not kid ourselves fundamentally about what's involved; *we cannot pretend to come in contact with madness, to manage it if need be, for long periods of time and remain wholly sane.* We should not kid ourselves: deterrence involves a fundamental Faustian bargain from which we have every reason to believe we have as much, if not more, to lose than we have to gain. The plain fact of the matter is that the presumption that we could remain "cool-headed" while using weapons of awesome anxiety is psychologically absurd. It goes against every grain of evidence that has been compiled in this century from the field of psychodynamics.

What then is the best we can hope for? Surprisingly, the field of psychodynamics teaches us that the answer is not the complete abandonment of all efforts directed towards deterrence. Even though, as I believe this essay has shown that traditional thinking about deterrence completely collapses because it rests on too many, shaky, dubious, and contradictory assumptions about human behavior to be fully credible any longer, psychodynamics does not preach that we abandon all efforts whatsoever aimed at the preservation of ourselves. To abandon all efforts of deterrence or protection would be in effect to abandon the existence of our egos. It would be in effect to permit our reality function to dissolve. Now if Freud and all the other great psychoanalysts taught us anything it is that there is a legitimate basis for the ego to wish to defend itself.

This is why I believe it is White [127] who more than anyone else hits the dilemma of deterrence on its head when he says we need a' la Freud to encourage "realistic" but not "exaggerated" fear in our adversaries. We need in short to couple a policy of minimum deterrence that fits our needs independently of the needs of the Soviets with drastic efforts at severe tension reduction. If all strategies of deterrence involve as I have continually stressed enormous contradictions and paradoxes, then it is a matter of picking your paradox and contradiction. Instead of managing massive amounts of rationality and irrationality and hoping for rationality, I would prefer instead to manage lesser amounts of weapons, even though I believe they can never be used because there is no guarantee that a nuclear war can remain limited once started, and to take significant and dramatic actions aimed towards tension reduction. There is at least the hope that with lesser amounts of weapons we can better contain a miscalculation. The dialectical counter to this (i.e., my own recommended policy is itself not free of contradiction and paradox) is that this policy may lead to overexaggerated self-confidence and in this sense lead to another "less" as well.

What this means in the end is that there will never be a strategy for managing nuclear weapons that does not contain significant paradoxical and contradictory elements from all the tables we have examined in this paper. This point is absolutely essential to understand if we are ever to hopefully contain our nuclear dilemma.

For this and other reasons, this essay has deliberately stayed away from debating the merits of particular proposals (formulas) for the reduction of specific weapon systems on both sides. It is not because these matters are unimportant that I have avoided this discussion. They obviously are of vital importance. Rather, I have tried to accomplish something much more basic. Particular formulas for reduction like particular weapon systems will come and go. What will not come and go if we have truly gotten to the heart of nuclear phenomena is the philosophical nature of the beast with which we are dealing. The particular formulas will never be any better, we will never make any progress in taming the nuclear dilemma until we realize that for every proposed benefit we envision there will inevitably be an equally strong cost associated as well. There is just no way out of this mess. It is inherent in the nature of the phenomenon with which we are dealing.

In this century, the fields of logic and economics have recognized two major contributions which have shaken their foundations to their very core. The first was a proof by the great logician Kurt *Gödel* that any system of logic rich enough to have significant theorems that could be proved within it would also contain other theorems that would contradict the others. It was impossible in short to have a logical system that was complete, i.e., contained within it every theorem that could be proved, and not have severe contradictions. One either had (a) completeness and contradictions or (b) incompleteness and no contradictions, but not (c) both completeness and no contradictions.

Similarly Kenneth Arrow proved within the field of mathematical economics that it was not possible to have a voting system that would satisfy all the desired conditions of a democracy, e.g., that there be a dictator, i.e., someone who was able to impose his voting preferences on everyone.

In a similar fashion, one would like to be able to prove conclusively that there will never be a theory of deterrence that will be free of significant paradoxes and contradictions. I don't believe we will ever be able to prove this formally although I can't prove that we won't be able to prove it either. Is this fundamental inability either to prove or to disprove itself a kind of uncertainty? Is it this kind of uncertainty that has kept the peace these thirty odd years? If it is, is this any kind of a guarantor for peace on which to stake the fate of the planet?

If as I have argued that the only reasonable course of action is to work towards minimal deterrence coupled with dramatic acts of tension reduction because these are the only courses of action of which I know that are consistent with the human mind's ability to manage realistic fear, how can one explain all of this to a wide public? How does one explain to a public that has been brought up and nourished by our political leaders, and even by experts who should know better, that "more no longer necessarily leads to more"? Is this the paradox that anyone must confront anyone who would explain a more reasonable policy to the American people?

How in short does one explain that the metaphysics of nuclear phenomena are not like the metaphysics of things with which we deal in everyday life? How does one explain any of this precisely because we are dealing fundamentally in a realm where things (policies, strategies, etc.) can no longer be proved conclusively one way or the other? We had better find a way for there is no more important thing to attempt. For what is true of our nuclear predicament (i.e., "more is less," "less is more") is now true of *every, single, one* of the major crises our so-called civilized world now faces: toxic waste, worldwide pollution, the energy crisis, worldwide inflation and stagnation.

Robert Jervis' excellent book, *The Illogic of American Nuclear Strategy*, [128] to which I have referred earlier, shows that of all the things the human mind is charged with accepting and with managing, paradox and contradiction are among the most difficult. Indeed, they are generally experienced as so intolerable that most people will attempt to escape from them through a series of devices that are by now so well-known that they can be explicitly catalogued: (1) the regaining of security through technological superiority; (2) persistent calls for total nuclear disarmament; (3) growing calls for freezing nuclear weapons at their current levels; and (4) persistent attempts to conventionalize nuclear arms, i.e., to treat them as weapons which are essentially no different from other kinds [129]. As an example of the first kind of attempt, Jervis cites President Reagan's desire to "regain the kind of security that was possible before 1945 by establishing an effective defense [i.e., what is commonly referred to as Star Wars]." [130] Since it is psychologically unbearable for those such as the President who see the Russians as "highly aggressive," if not the very epitome of evil, to live in the kind of world in which our two fates are inseparably linked together because of the sheer levels of nuclear weapons and the magnitude of their destructive effects on both our civilizations, it is vital that we break the deadlock and regain our superiority.

What Jervis correctly points out is that tactics (1) and (4) for breaking the nuclear deadlock in which both the US and USSR find themselves are now impossible in the kind of complex world we have constructed. While few would argue for stopping technological development and would completely deny the possibility of a technological breakthrough ever occurring by one side, it now appears impossible that one side or the other will pull dramatically ahead of the other, remain ahead for long, or know how to exploit such advantages even if they could pull ahead and remain there. As for tactics (2) and (3), as worthy as they are—and make no mistake they are worthy—they still ignore the fact that the paradoxes and contradictions I have discussed in this article are here to stay. They are a permanent feature of the kind of highly interconnected civilization we have constructed. For example, even if all nuclear weapons were removed from the face of the planet, we would still be left with an infrastructure that is more susceptible to widespread disruption by terrorist activity than was possible in earlier, simpler ages. For instance, it is estimated that 80 percent of the liquid natural gas in the U.S. flows through the single state of Louisiana and that therefore a single terrorist with a hundred pounds of dynamite strategically placed could therefore shut down this entire country [131]. One doesn't therefore need nuclear weapons to disrupt our civilizations when something seemingly much more "safer" (1) will do the trick.

However, because such paradoxes and contradictions will not easily or quickly go away is not therefore an argument for the continued existence of nuclear weapons at their current bloated levels. If anything, this entire article is an argument for their severe reduction if only for the reason that we hope we can more reasonably contain and manage the paradoxes they unleash.

In his book, *Redesigning the Future* [132], Russell L. Ackoff outlines four radically distinct postures towards planning with regard to the future. They are: inactivism, reactivism, preactivism, interactivism. Inactivism is the posture that recommends doing nothing. It says that problems will go away if we simply ignore them. The only kinds of organizations that have the luxury of doing this are those that are protected from their environment. Traditionally this has been universities and government agencies, although as both have found in recent years they are less protected from external competition than ever before. There are more external private sources of competition than schools or government have ever had to face.

Given the widespread fear, anxiety, and helplessness that the sheer existence of nuclear weapons set off, plus the fact that I believe that most people subliminally recognize the existence of the paradoxes and contradictions I have been talking about even if they do not explicitly acknowledge or know how to deal with them, why should we be surprised to learn that most people instinctively practice an unwilling inactivism? That is, they believe that they or no one can really do anything to deal with the nuclear mega-mess in which we find ourselves.

Reactivism is the philosophy of going back to the tried and true ways that worked so well in the past. Reactivism is also just what its name implies. It advocates taking action after something has occurred, not before. Is it therefore not clear that the first and fourth tactics that Jervis identified for avoiding paradoxes and contradictions are essentially reactive, and as such, fail because they assume that what worked for a simpler, disconnected world can work for a systemic, highly interconnected one?

Preactivism is just the opposite of reactivism. It postulates the great benefits to be obtained from the great leap ahead into a desirable future without specifying in detail the operational steps for getting from "horrible here" to "terrific there." Is it not clear that Jervis' tactics (2) and (3) are essentially of this kind?

What's left? What Ackoff calls interactivism of the philosophy that recognizes that there is a valid element in each of the prior four planning outlooks, but not necessarily in the original ways in which they have been put forth. Thus, for example, I have no difficulty in accepting the idea of a nuclear freeze but if and only if it is clearly recognized that the deep paradoxes and contradictions which haunt out nuclear dilemma will never vanish unless we create a wholly new kind of civilization which goes drastically beyond that which we have now.

Over half a century ago, John Dewey, one of America's greatest philosophers, wrote an incredible book. It was entitled *The Quest for Certainty* [133]. Dewey describes how Western man's entire search for knowledge was founded on a fundamental presumption: that certainty was both possible and desirable in human affairs; indeed, without certainty one could not be said to have obtained knowledge. No matter that radically competing brands of Western philosophy differed sharply in what they located as the source of certainty; they all accepted the same basic premise that certainty in and of itself was as inherently desirable as it was necessary.

Dewey argued persuasively that the quest for certainty was an illusion, that it was doomed to failure, that in effect it was neither desirable nor possible to achieve in human affairs. It was not that Dewey thereby expected men to give up their quest for certainty, a quest he labeled neurotic, for he understood well the source of this quest. It was rooted deep in man's compulsive need for order against a world that he often found chaotic. In a sense, Dewey thus anticipated some 56 years earlier every one of the principal conclusions of this essay. What he could not anticipate, but I sense he would not be surprised to find, that the basic need for certainty is wrapped up with the most vital and global issue of our time. Were he still alive today, I believe he would be cautioning us more forcefully than ever to give up our false illusions that somehow, somewhere out there is a way of comprehending our nuclear dilemma that is free of paradox and contradiction, and in short we can manage the biggest dilemma of all on a bedrock of perfect certainty.

5 PRINCIPAL CONCLUSIONS

1. The credibility for any and all theories of nuclear strategy and deterrence rest on the shakiest of grounds. They rest on too many dubious, unproved, and even unprovable assumptions about human behavior to provide any reliable guide for action and for feelings of security any longer.
2. It is highly doubtful that nuclear weapons could be reliably and/or controllably used under any scenario. There are no firm, reliable, or valid reasons whatsoever for believing that one could control under the heat of battle what one was unable to control before the onset of battle. It is hard in other words to believe that constraint and rationality would suddenly and magically appear in the midst of unconstraint and irrationality.
3. If the human mind seems to need terrifying calamities or the potential for them "to remind itself of the perpetual need to restrain itself," is the best we can do is to minimize this potential, if we cannot fully eliminate it? If so, does this argue for minimal deterrence as it has been proposed by White, although this like all positions will itself not be free of contradictions and paradoxes?
4. As described by Strobe Talbott, there seems to be little if any recognition within the Reagan administration, and to be fair by the vast majority of those opposing forces outside of the administration, of the conditions described in this article which drastically undermine all current theories of nuclear strategy. As a result, there is and will continue to be in the unforeseen future, a vast, unbridgeable gap between (a) the existence of large numbers of nuclear weapons for essentially political (symbolic) reasons and (b) the no longer intellectually supportable reasons for their continued existence on philosophical grounds.
5. If the grounds for the philosophical collapse of conventional theories of nuclear deterrence as presented here are valid, then it will require the most immense and intense acts of political courage—better yet, heroism—to convince the American people that they would not be giving away any vital security for them to cut

back drastically and unilaterally on their own nuclear arsenal. If, for instance, theories of nuclear winter should "prove correct," and I think that they will, then of what good are weapons whose use against one's enemies proves to be just as suicidal against one's own population? But if this is the case, then we first need to cut our own levels of weapons below that which if used would initiate a nuclear winter upon ourselves. Our bargaining is thus as much with ourselves as with our enemies. The US and the USSR may both have to unilaterally cut their weapons below the level needed to initiate a nuclear winter before they can seriously bargain to reduce their levels even further.

6. "More is no longer more." Even our worst and most determined adversaries recognize this viscerally even if they do not always understand why this is so intellectually. We have no choice but to learn to think very differently if we are to manage the biggest mess that has been dumped on the human race.

REFERENCES

1. This essay was completed before I had the opportunity to read Robert Jervis' excellent and provocative book, *The Illogic of American Nuclear Strategy*, Ithaca: Cornell, 1984. Although it comes tantalizingly close to the position advocated here, it does not carry the notion of paradox and contradiction as deeply and as pervasively as I argue in this essay. What Jervis calls the "tension" inherent in traditional notions of deterrence I believe is better captured by the concept of dialectical contradiction. The position advocated here is thus much more extreme than his. This does not detract from either the cogency of his arguments or the overall excellence of his book.
2. R. O. Mason and I. I. Mitroff, *Challenging Strategic Planning Assumptions*, New York: John Wiley, 1981; I. I. Mitroff, R. O. Mason, and V. P. Barabba, *The 1980 Census: Policy Making Amid Turbulence*, Lexington, MA: Lexington Books, 1983.
3. Ralph K. White, *Fearful Warriors, A Psychological Profile of US-Soviet Relations*, New York: The Free Press, 1984, p. 138.
4. Fred Kaplan, *The Wizards of Armageddon*, New York: Simon and Schuster, 1983, p. 378.
5. Daniel Frei, *Risks of Unintentional Nuclear War*, Totowa, New Jersey: Rowmand & Allanheld, 1983, p. 98.
6. Kaplan, *op. cit.*, p. 341.
7. Richard K. Betts, "Nuclear Weapons," in Joseph S. Nye, Jr. Ed., *The Making of America's Soviet Policy*, New Haven, CT: Yale, 1984, p. 117.
8. Frei, *op. cit.*, p. 66.
9. Kaplan, *op. cit.*, p. 234.
10. Strobe Talbott, *Deadly Gambits*, New York: Knopf, 1984, p. 66.
11. Frei, *op. cit.*, p. 205; Michael Sheehan, *The Arms Race*, Oxford: Martin Robertson, 1983, pp. 200-201.
12. Helen Caldicott, *Missile Envy, The Arms Race and Nuclear War*, New York: Morrow, 1984, p. 163; Frei, *op. cit.*, p. 162.
13. Caldicott, *op. cit.*, p. 163.
14. Patrick Glynn, "The Moral Case for the Arms Buildup," in R. James Woolsey, ed., *Nuclear Arms, Ethics, Strategy, Politics*, San Francisco: ICS Press, 1984, p. 45.
15. Robert Kupperman, "Vulnerable America," in R. James Woolsey, Ed., *Nuclear Arms*, San Francisco: ICS, 1984, pp. 195-196.
16. Frei, *op. cit.*, p. 225.
17. Thomas C. Schelling, *The Strategy of Conflict*, New York: Galaxy, 1963, p. 122.
18. Sheehan, *op. cit.*, p. 83.
19. Frei, *op. cit.*, p. 90.

20. Kaplan, *op. cit.*, p. 390.
21. *Ibid.*, p. 347.
22. Stanley Hoffman, "Detente," in Joseph S. Nye, Jr., Ed., *The Making of America's Soviet Policy*, New Haven: Yale, 1984, p. 245.
23. Marshall Goldman and Raymond Vernon, "Economic Relations," in Nye, *op. cit.*, p. 168.
24. Alexander I. George, "Political Crises," in Nye, *op. cit.*, p. 125.
25. Richard K. Betts, "Nuclear Weapons," in Nye, *op. cit.*, p. 109.
26. Charles Krauthammer, "On Nuclear Morality," in R. James Woolsey, Ed., *Nuclear Arms, Ethics, Strategies, Politics*, San Francisco: ICS, 1984, p. 13.
27. Kaplan, *op. cit.*, p. 371.
28. Schelling, *op. cit.*, p. 193.
29. Brent Scowcroft, "Understanding the US Strategic Arsenal," in Woolsey, *op. cit.*, p. 76; Sheehan, *op. cit.*, p. 30.
30. Frei, *op. cit.*, p. 178-179.
31. Frei, *op. cit.*, p. 177.
32. Colin S. Gray, "Arms Control: Problems," in Woolsey, *op. cit.*, p. 159.
33. *Ibid.*, p. 163.
34. *Ibid.*, p. 166.
35. Fallows, *op. cit.*, p. 169.
36. W. J. Hannon, "Seismic Verification of a Comprehensive Test-Ban," *Science*, Vol. 227, No. 4684, January 18, 1985, p. 256.
37. Barrie Paskins, "Proliferation and the Nature of Deterrence," in Nigel Blake and K. Pole, Eds., *Dangers of Deterrence, Philosophers and Nuclear Strategy*, London: Routledge and Kanpual, 1983, p. 128.
38. William J. Eaton, "Gromyko Vows to Fight Space Arms," *Los Angeles Times*, January 14, 1985, p. 10.
39. Jervis, *op. cit.*, p. 44.
40. *Ibid.*, p. 62.
41. Russell L. Ackoff, *Creating the Corporate Future, Plan or Be Planned For*, New York: Wiley, 1981.
42. Fallows, *op.cit.*, p. 147-148.
43. Schelling, *op.cit.*, p. 192.
44. Kaplan, *op.cit.*, p. 260.
45. Schelling, *op.cit.*, p. 18, 22.
46. Frei, *op.cit.*, p. 89-90.
47. Schelling, *op.cit.*, p. 40.
48. White, *op.cit.*, p. 53, 114.
49. Schelling, *op.cit.*, p. 79.
50. George, *op.cit.*, p. 131.
51. White, *op.cit.*, p. 74.
52. White, *op.cit.*, p. 80.
53. *Ibid.*, p. 307.
54. Schelling, *op.cit.*, p. 234.
55. Kaplan, *op.cit.*, p. 331.
56. Talbott, *op.cit.*, p. 34.
57. *Ibid.*, p. 254-258.
58. Frei, *op.cit.*, p. 104.
59. Sheehan, *op.cit.*, p. 153.
60. Frei, *op.cit.*, p. 29.
61. Sheehan, *op.cit.*, pp. 65, 82.
62. David Holloway, *The Soviet Union and the Arms Race*, New Haven: Yale, 1983, p. 152.
63. *Ibid.*, p. 107.

64. Betts, *op.cit.*, p. 114.
65. Talbott, *op.cit.*, p. 102.
66. Sheehan, *op.cit.*, p. 32.
67. Kaplan, *op.cit.*, p. 192.
68. Frei, *op.cit.*, p. 33.
69. Krauthammer, *op.cit.*, p. 15; Michael Quinlan, "Thinking Deterrence Through," in R. James Woolsey, *op.cit.*, p. 62.
70. White, *op.cit.*, p. 104.
71. Kaplan, *op.cit.*, p. 167.
72. Talbott, *op.cit.*, p. 124.
73. *Ibid.*, p. 23.
74. Hoffman, *op.cit.*, p. 248.
75. Talbott, *op.cit.*, p. 100.
76. Sheehan, *op.cit.*, p. 192.
77. *Ibid.*, p. 99.
78. Talbott, *op.cit.*, p. 309.
79. Frei, *op.cit.*, p. 106.
80. Holloway, *op.cit.*, p. 104.
81. *Ibid.*, p. 54.
82. Frei, *op.cit.*, pp. 171-172.
83. Betts, *op.cit.*, p. 111.
84. White, *op.cit.*, p. 118.
85. Amory B. Lovins and L. Hunter Lovins, "Reducing Vulnerability: The Energy Jugular," in Woolsey, *op.cit.*, p. 192.
86. Joseph S. Nye, Jr., "Can America Manage Its Soviet Policy," in Nye, *op.cit.*, p. 330.
87. Sheehan, *op.cit.*, p. 156.
88. *Ibid.*, p. 55.
89. White, *op.cit.*, p. 33.
90. *Ibid.*, p. 69.
91. *Ibid.*, p. 72.
92. Hoffman, *op.cit.*, p. 248.
93. Holloway, *op.cit.*, p. 20.
94. Frei, *op.cit.*, p. 107.
95. Holloway, p. 72.
96. Sheehan, *op.cit.*, p. 43.
97. Krauthammer, *op.cit.*, p. 21.
98. Krauthammer, *op.cit.*, p. 21.
99. Frei, *op.cit.*, p. 8.
100. Quinlan, *op.cit.*, p. 61.
101. Schelling, *op.cit.*, p. 238.
102. Holloway, *op.cit.*, p. 164-165.
103. Talbott, *op.cit.*, p. 33.
104. Betts, *op.cit.*, p. 102.
105. Patrick Glynn, "The Moral Case for the Arms Buildup," in Woolsey, *op.cit.*, p. 31.
106. *Ibid.*, p. 36.
107. *Ibid.*, p. 46.
108. *Ibid.*, p. 47.
109. Quinlan, *op.cit.*, p. 57.
110. Betts, *op.cit.*, p. 97.
111. Sheehan, *op.cit.*, p. 111.
112. Frei, *op.cit.*, p. 84.
113. Kaplan, *op.cit.*, p. 191.

114. Sheehan, *op.cit.*, p. 16.
115. Krauthammer, *op.cit.*, p. 14.
116. Glynn, *op.cit.*, p. 28.
117. *Ibid.*, p. 30.
118. Sheehan, *op.cit.*, p. 186.
119. *Ibid.*, p. 192.
120. White, *op.cit.*, p. 75.
121. Frei, *op.cit.*, p. 85.
122. Kaplan, *op.cit.*, pp. 289-301.
123. *Ibid.*, p. 327.
124. *Ibid*, p. 274.
125. White, *op.cit.*, p. 87.
126. Talbott, *op.cit.*, p. 127.
127. White, *op. cit.*, pp. 53, 114.
128. Jervis, *op. cit.*
129. *Ibid.*, pp. 54-58.
130. *Ibid.*, p. 54.
131. Lovins and Lovins, *op. cit.*, p. 182.
132. Russell L. Ackoff, *Redesigning the Future*, New York: Wiley, 1974.
133. John Dewey, *The Quest for Certainty*, New York: G. P. Putnam, 1929.

DISCUSSION OF MITROFF'S PRESENTATION

Sadovsky

May I ask you about the nature of your paradoxical situations? Are they logical ones? It seems to me that not all of them are logical. Are they epistemological, social etc.? From a knowledge of the nature of paradoxical situations perhaps we can find ways of resolving them.

Mitroff

There are nuclear paradoxes which fall into every type. Indeed, this is why the recognition of paradox is so important.

Von Glasersfeld

Would you agree that some of your "paradoxes" can be resolved on the level of semantics? For example, weapons are things that kill your enemy and are not supposed to kill you yourself. Thermonuclear bombs—a small number of them—will kill the planet. Hence, they should not be called "weapons", "arms", "defense", etc.

Mitroff

Only a small part can be resolved through semantics for no matter how they are worded, the essential nature of the paradoxes still comes through. The reason is that the phenomenon fits into all four major types no matter what the wording. This very "looseness" of words is itself one of the characteristic features of complex problems. The difficulties arise in part because of their global nature. They cut across narrow attempts to contain them.

Apokin

Have you tried to apply game theory to the problem of international negotiations and to predict the results?

Mitroff

No.

Pospelov

Rapoport and Chamach conducted some experiments that involved modeling human behavior in "prisoner's dilemma" games. They investigated the following game matrix.

	first player	
second player	(5; 5)	(-10;10)
	(10;-10)	(-1; -1)

In the presence of information about the choice of a strategy, people select the first strategy and always win 5 units (the point (5;5)). But in the absence of information on the choice, the stable point becomes point (-1;-1) where both players lose. This is due to the absence of altruism and lack of trust in one another.

This experiment shows the importance of information exchange for eliminating the situation in which "worse is better because it is less trouble." It seems that information can prevent some paradoxical situations.

What amount of information on the number of system states, the number of subsystems and the number of arguments about each subsystem is needed for reconstruction in a specific case?

Lapin

Assuming that the paradoxes of nuclear policy are well known at the highest governmental levels. How would this knowledge contribute to resolving the paradoxes?

Mitroff

First, I do not believe that there is such knowledge at the highest levels of our government, but if there were, then it would make clear the fallacy of either side's trying to prevail through technology alone. It would show that once both sides have a lethal strike capability, then neither can pull dramatically ahead of the other.

Lavrov

The history of science has given us some wonderful examples. Suppose we have constructed a formal theory for some concrete model. If we go back, we can build a new model which is unlike the original model, but it satisfies our formal theory. It is a nonstandard model. I think we must often look for nonstandard solutions to our social problems.

CHAPTER

4

CONCEPTUAL DIFFICULTIES INVOLVED IN RELATIONS BETWEEN THE UNITED STATES AND THE USSR

Stuart A. Umpleby

The problem of devising a successful strategy for reducing conflict between the United States and the Soviet Union has resisted solution for several decades. Assuming that the difficulty may lie in our conceptualizations, three schools of thought within American cybernetics and general systems theory are examined to determine how each of them would approach the problem. First the arms race and then the tension that generates the arms race are considered. The conclusion is that additional work on underlying conceptual systems would be helpful and that promising theoretical approaches already exist to guide this work.

1 INTRODUCTION

The effort to create theories of social systems encounters certain logical difficulties which do not arise quite so readily when creating theories in the natural sciences. This paper explores some of these issues as viewed by three schools of thought within the cybernetics and systems community in the United States. The focus of attention will be on relations between the United States and the USSR in the hope that these explorations may help to lay the foundation for greater mutual understanding between the people in these two countries in the years ahead. The three schools of thought are cybernetics, the systems approach, and reflexive structures.

Cyberneticians working in the West in the 1970's and 1980's, including Ross Ashby, Heinz von Foerster, Humberto Maturana, Ernst von Glasersfeld, and Lars Löfgren, were trying to devise deductive theories of adaptation, cognition, and self-regulation. They were quite familiar with Gödel's incompleteness theorem and the paradoxes that result from self-reference (Nagel and Newman, 1958).

A second group, usually associated with the systems approach, includes C. West Churchman, Russell Ackoff, Ian Mitroff, and other followers of the philosopher E. A. Singer, Jr. This group has been interested in purposeful systems and has focused on developing methods to aid decision-makers. Contrary to most academics, this group has been skeptical of efforts to build theories, or at least disciplines. They favor inquiry which is guided by a primary concern with implementation.

The third body of work is that initiated by Vladimir Lefebvre (1982) who has developed a theory of two types of ethical cognition. This theory is particularly relevant to a meeting involving American and Soviet scientists.

In his paper in this volume Ian Mitroff calls attention to the paradoxes involved in strategic thinking. I choose to interpret his paper as raising the question of whether the intractability of the tension between the United States and the Soviet Union is due not just to rivalry between two superpowers but rather to some unusual logical difficulties. Hence I propose to look seriously at the logics involved in East-West communication. I shall begin with a review of the origin of some types of paradox.

2 HOW SELF-REFERENCE CAN LEAD TO PARADOX

Conflict is not the same as paradox, which the American Heritage Dictionary defines as follows:

> Paradox. 1. A seemingly contradictory statement that may nonetheless be true. 2. A person, situation, or action exhibiting inexplicable or contradictory aspects. 3. An assertion that is essentially self-contradictory, although based on a valid deduction from acceptable premises. 4. A statement contrary to received opinion.

The cyberneticians referred to earlier have dealt with paradox as a consequence of self-reference (Umpleby, 1981). At the turn of the century Bertrand Russell (1903) recognized that sets which contain themselves will lead to paradox. A frequently used example is the village barber. Imagine a village consisting only of men. There are two kinds of men—those who shave themselves and those who are shaved by the barber. The question is, who shaves the barber? If he shaves himself, then he is shaved by the barber. But if he is shaved by the barber, then he shaves himself. If we assume that he is in one set, then it turns out that he is in the other and vice versa. Any time we formulate a set which refers to itself, we will generate a paradox.

Another example is Epimenides' Paradox. A man steps off a boat from Crete and announces, "All Cretans are liars." If he is telling the truth, then he is lying. But if he is lying, then he is telling the truth. If the man from Crete were to say, "All

Cretans except me are liars," then there would be no paradox. But as soon as a set contains itself, a paradox results.

Self-reference is of more than merely academic interest. Three quite practical examples will show why.

The first example of self-reference involves problem formulation. Definitions of problems reveal at least as much about the observer as about the situation observed. During a recent war a particular field hospital had not reported any patients requiring psychological help, until a doctor was assigned there who had had psychological training. He reported that there were many psychological cases and requested that a professional psychologist be assigned to the hospital. Problems can remain unsolved simply because a person with the needed knowledge is not present. Or, a definition of a "problem" may say more about the observer than about the situation observed. People tend to define problems that they know how to solve.

A second example of self-reference involves theories of social systems. A social scientist creating a theory of society is an element of the society which he is describing. We might then ask whether it is possible for a scientist to create an unbiased description of the society or whether he will inevitably describe the society in a way that will appeal to the members of a group in which he would like to be accepted. In what sense could a theory of society be objective or are all social theories by necessity self-serving?

A third example of self-reference relates to artificial intelligence. If we are trying to create an intelligent machine, one of the things that we would like for the machine to be able to do is to talk about itself. That is, it should be able to construct sentences such as the following: "I believe...", "I remember...", or "My opinion is...". Such a machine would require a metalanguage in order to converse about the language in which "thought" is taking place. But we would also like the machine to be able to converse about the metalanguage, that is, to construct sentences such as, "I believe X because..." and "I find it easiest to remember X when..." This level of discourse would require a metametalanguage. The analysis could be continued as we require higher and higher levels of self-reflection in our machine. If we use the theory of a hierarchy of types (Russell, 1903), it would seem that such a machine would have to be infinitely large. But we have an existence proof that such a large machine is not required, namely the human brain. Somehow the human brain manages to produce at least several levels of self-reflection without being very large.

Our preference in designing machines is to first produce a logically consistent theory of the phenomenon we are trying to produce. The problem of self-reference seemed to confound the possibility of creating a paradox-free formal theory of cognition. Recent work has led researchers to believe that the problem lies not in the phenomenon itself but rather in the mathematical language used to describe the phenomenon. What is needed is a calculus which permits paradox rather than *a priori* forbids it. Rather than using a space-based mathematics such as set theory, a calculus such as category theory, which incorporates an implied time dimension, is more appropriate (Arbib and Manes, 1975).

With this background on the problem of self-reference, the paradoxes it generates, and the language in which it is discussed, let us now return to the issue of strategic thinking.

3 LEVELS OF THEORIZING IN THE FIELD OF NUCLEAR STRATEGY

As Mitroff points out in his paper in this volume, the field of nuclear strategy is rife with logical paradoxes. These result in part from the fact that there are many different strategies functioning on a variety of levels of analysis. A strategy that makes sense on one level is often counter-productive on another level. Four levels of analysis will illustrate what I have in mind.

1. Technology or weaponry. Whether the US builds Pershing missiles, cruise missiles, submarine launched missiles, missiles in hardened silos, mobile land missiles, or a space-based defense system is in part determined by what current technology is able to produce.

2. Military strategy. Current NATO strategy is to counter a land invasion of Western Europe with the first use of nuclear weapons. This strategy has been chosen because it is less expensive. It is cheaper to make nuclear weapons than to maintain a large army in Western Europe. The Soviet strategy in Europe is to preempt the NATO attack.

3. Psychological strategy. The U.S. has said that it was introducing Pershing missiles into Europe as a result of Soviet installation of SS 20 missiles. The U.S. move was presented as a response to a Soviet move. At this level the installation of a new weapon is an event in a psychological and political battle (Axelrod, 1984).

4. Social system strategy. Another theory holds that the U.S. should take advantage of the fact that it has a stronger economy (Possony and Pournelle, 1970). According to this strategy the U.S. should escalate the arms race in the hope of either achieving an advantage in weapons or of increasing dissatisfaction within the Soviet Union if the Soviet government tries to keep pace by taking resources away from its domestic economy.

If attention is focused on level 2, where most "military thinking" occurs, "rationality" may seem to prevail. However, nuclear strategy can become quite paradoxical at level three which emphasizes psychological factors. For example, Pershing missiles are vulnerable to a Soviet first strike. It has been suggested that their vulnerability increases instability in Europe. This occurs because it is easier, and hence more likely, for the Soviet Union to launch a preemptive strike against land-based Pershing missiles than against submarine-based missiles. Hence, an effort to restore the balance of power in land based missiles in Europe may actually reduce stability.

But there is a second paradox involving the vulnerability of the Pershing missiles. If one implication of the decision to install Pershing II's is to lower the threshold of a preemptive strike, a second result is that political stability might increase. That is, if the threshold of war is lowered, the desire to avoid disturbing the peace would increase. If both sides recognize the greater danger, both sides will be more reluctant to risk a confrontation.

There is a further paradox involved in the psychological approach. If both sides take a tit-for-tat approach to weapons deployment, there is no logical limit to how many weapons can be deployed. The idea in this strategy is to match each move by the opponent. The doctrine of deterrence, on the other hand, emphasizes weapons which can survive a first strike and hence convince an aggressor that his country would suffer catastrophic retaliation. If one's strategy is deterrence, one can achieve sufficiency in weaponry. However, if the deployment of weapons is viewed as a game for psychological or propaganda effect, the number of weapons that could be deployed is unlimited. Hence, an effort to "teach an opponent a lesson" when occurring in the context of changing technology, can lead to an arms race with no logical limit to the number of weapons.

Two other factors should probably be mentioned. First, there are institutional considerations such as the triad strategy with its roots in the organization of the Defense Department and interservice rivalry. The organization of the Defense Department can lead to more weapons being deployed than are necessary for deterrence simply in order to give each branch of the military "a piece of the strategic pie." Other institutional considerations are the defense industries and the Congressmen from key defense districts who want to keep jobs in their districts regardless of the strategic or international political implications of the products which those jobs are associated with.

4 WHY DO THESE INCONSISTENCIES ARISE?

As these examples show, it is possible to have conflict between levels of thinking. But this case has not been thoroughly analyzed within the systems field. More often compatibility between levels of analysis is assumed. For example, James G. Miller in his book *Living Systems* describes seven levels of systems (i.e. cell, organ, organism, group, firm, nation, world). Systems at each level are assumed to possess nineteen critical subsystems (e.g. ingester, distributor, decoder, memory). Miller tends to assume that these levels are or can be in harmony with each other. Activities at a lower level are not only compatible with the next higher level but in fact make possible the next higher level. The general systems theory literature has tended to emphasize similarities and compatibilities among levels rather than dissimilarities or incompatibilities.

In the example of strategic thinking such delightful harmony does not exist. A major difference between Miller's living systems theory and strategic thinking is that Miller was talking about tangible systems. Although strategy is concerned with the movement of men and weapons, strategic thinking involves levels of analysis which are basically conceptual. This is an important difference, but by itself it does not explain the harmony among levels in living systems theory and most general systems theorizing and the disharmony among levels in strategic thinking. I believe it is fruitful to identify the reasons for the disharmony in strategic thinking.

First, there can be a conflict between short-term considerations and long-term considerations. For example, installing a new weapons systems may have a brief near-term value as a gesture showing resolve even though its long-term strategic utility may be questionable. Also, money spent now is not available to spend later.

Second, there can be conflicts between points of view because of institutional rivalries and budget limitations. For example, submarine-launched missiles mean more money for the Navy while land-based missiles mean more money for the Army. A final decision will have to serve domestic constituencies as well as short and long term strategic considerations. It is quite unlikely that a logical analysis of a complex policy issue will produce a conclusion that is consistent on all levels and between levels and also will be one that all individuals and groups will support.

But there is more going on in the case of strategic thinking than just disagreements on the allocation of resources either in time or among groups. Strategic thinking is not a normal policy issue. The difficulties involved are not simply those of allocating finite resources. Two other difficulties relating to self-reference are involved.

First, there is the certainty that each party is lying to some degree. Given an international rivalry that might lead to war, it is reasonable to presume that each side will try to mislead the other on some subjects. Just as in the case of the person from Crete, not all statements or actions from the other side can be taken at face value, simply because they come from the other side.

Second, threats to take actions that would destroy one's own country as well as an opponent's are not inherently believable. This is the source of many of the paradoxes in nuclear strategy—the fact that our strategy for maintaining our way of life involves threatening to use nuclear weapons, when a nuclear war might result in a nuclear winter which could destroy civilization as we know it. Hence, nuclear strategy is based on our ability to convince an adversary that we are willing to risk destroying what we are attempting to save. "Rational" strategies are possible only if one accepts the "rationality" of this starting point.

5 IS A THEORY OF STRATEGIC THINKING POSSIBLE?

Professional strategists tend to focus on one level of analysis. At each level there are theories that explain cause and effect relationships. Scientific theories usually operate on only one level of analysis. Not only disciplines but also universities tend to be structured in this way. Universities, at least in the West, have departments of psychology, sociology, business and political science. Decision-making by individuals, families, organizations, and nations is thus divided largely according to how many people are involved in the process. Theories that link levels of analysis do occur but usually in applied fields such as management. For example, the effect of personalities on the behavior of a group is a subject of study in the field of management. Another theme of interest within management is how organizational policies affect the morale and motivation of employees.

In practice we routinely use several models or levels of analysis when thinking about social systems. However, the common conception of theory is limited to one or at most two levels of analysis. This gap between the richness of everyday thought and the more limited conceptions of science has led some system scholars to turn away from efforts to develop theories. I believe, however, that there is another choice. Either we can choose not to do theoretical work, where theory is defined in the traditional way, or we can change our conception of theory. Churchman, Mitroff and others seem to me to have chosen the first alternative.

If one assumes that theories operate on one or a few levels, yet one also observes that there are several different levels of theories involving nuclear strategy, and, furthermore, one notes the large number of paradoxes that these various theories generate, then one might, indeed, become discouraged with the prospect that theory could lead us out of our difficulty.

However, suppose we make a distinction between cybernetics as a theory of regulation and other theories as illuminating some aspect of the system being regulated. We could furthermore note that any theory is a reduction of the phenomenon being studied. Each theory (including cybernetics) illuminates only part of the puzzle. A decision-maker would then have to weigh each theory and the considerations it focuses on, consider them in the light of his overall objectives and render some final judgment. It may not be possible to develop a single, integrated, logically consistent theory of the phenomenon under investigation, but that does not mean that it is impossible to develop a formal theory of regulation, or that work on the development of such a theory is not useful. A theory of regulation would include systems that regulate themselves and hence would include self-reference.

6 ALTERNATIVE EXPLANATIONS OF EAST-WEST TENSION

The United States and the Soviet Union have been engaged in a "cold war" since soon after the end of World War II. Although the cold war may not be either rational or stable, we can at least be thankful that the tension has not escalated to a nuclear war. Nevertheless, the danger inherent in the existence of nuclear weapons and the very large sums of money that are devoted to military activities suggest that a change in the current pattern of behavior would be to everyone's advantage. Efforts to control military spending and to reduce risk through arms control have not been notably successful. It may be that the time has come to consider alternative strategies. That is, rather than trying to limit the arms race, we might focus our attention on why there is such intense distrust between the United States and the Soviet Union. Several explanations of the current conflict have been proposed.

Cultural differences are one explanation. People in general tend to believe that their behavior is reasonable, appropriate and good. Hence cultural differences often lead to conflict because behavior which is different frequently is interpreted as being bad. Yet cultural differences alone rarely lead to arms races. If cultural differences explain the tension between the United States and the Soviet Union, they must be either unusually strong cultural differences, or cultural differences of a different kind.

A second explanation for the arms race is that it is the modern version of nations seeking to establish "spheres of influence." According to this view each large nation state attempts to surround itself with allies and to compete with other large powers for allies. Yet this kind of behavior is anachronistic in a world that since World War II has been moving from political and military competition to economic competition.

A third way of explaining East-West conflict is to assert that the military and industrial establishments in both countries have an interest in continued hostility and military preparations. Although institutional interests are undoubtedly a factor, any institution functions within a larger society. If an organization, such as the military, is thought to be less necessary than it once was, its share of the budget will be reduced.

The ideologies of the two countries and the assumption that these are incompatible is a fourth explanation of the current conflict. The idea of class struggle is still appealing to many people, but it explains conflict within a country better than conflict between two superpowers. These explanations of the conflict between the United States and the Soviet Union are well known and yet the intensity of the rivalry between the two countries continues to be puzzling to many people.

7 A NEW EXPLANATION FOR EAST-WEST TENSION

A new explanation of the conflict between the West and the Soviet Union has been offered by Vladimir Lefebvre (1982). He suggests that there are two fundamentally different systems of ethical cognition and that one is characteristic of the West while the other is characteristic of the Soviet Union. Lefebvre maintains that people in the West and in the Soviet Union not only value different things but also have different ways of valuing. Understanding his theory requires understanding how he uses the words "compromise" and "conflict". Compromise and conflict can occur both within an individual and between individuals. His concept of compromise and conflict between individuals corresponds to normal usage, but his concept of compromise and conflict within an individual is unusual.

To understand how Lefebvre thinks of compromise and conflict within an individual, think of a conflict between means and ends. If there is a conflict between means and ends, a person in a society in which the first ethical system dominates is considered to be a good person if he thinks about that conflict and takes it seriously. In a society which primarily uses the second ethical system, a person is considered to be a good person if he minimizes or compromises on a difference between means and ends. According to the second ethical system a good person is one who will pursue a good end even if the means are bad, whereas under the first ethical system a good person is one who will not use a bad means to achieve a good end. The constitutions of the United States and the Soviet Union reflect these two ways of thinking. The Constitution of the United States lists a series of "thou shalt nots." The government is not permitted to infringe freedom of speech, freedom of the press, the right to assemble, etc. The Constitution of the Soviet Union says that the government will guarantee housing, education, and health care, presumably by any means necessary.

As in the theories cited earlier, a paradox occurs in Lefebvre's theory as well. A person who compromises within, overlooking conflict between means and ends, tends to engage in conflict with other people. However, a person who engages in conflict within regarding means and ends tends to be willing to compromise with other people. That is, a person who is not reflective and who suppresses doubts about a conflict between means and ends, is likely to be uncompromising with other people. On the other hand, a person who is concerned about what is an appropriate means and who worries about whether some means are justified is likely to be more tolerant of others and hence less likely to be in conflict with others.

8 SOME CONSEQUENCES OF THE DIFFERENT ETHICAL SYSTEMS

These different ways of thinking can lead to different ways of negotiating. When a Soviet person begins a negotiation, he often leads off with a confrontational statement in an effort to present himself as a person of high ethical status who will not compromise his principles. An American will often begin a negotiation with a conciliatory statement intending to present himself as a person of high ethical status who is willing to give a little in order to get a little. Unfortunately each opening statement establishes exactly the opposite impression as the one intended. Americans come away thinking that the Soviets are boorish, aggressive, and unwilling to negotiate seriously. The Soviets conclude that the Americans are weak, unprincipled, and unprepared to negotiate seriously. With these unintended first impressions clearly established, the negotiation then proceeds.

Not only do Americans and Soviets have different ways of initiating a negotiation, different feelings are also involved. Americans approach a negotiation by indicating the area in which an agreement might be possible. They expect the other side to do the same. The region of overlap then becomes the focus of attention. The rest of the negotiation is envisioned as a process of clarification, wording, and working out of details. At the end of the process the negotiators are left with the satisfying feeling that they have jointly constructed a mutually acceptable agreement. This cooperatively produced success is interpreted as a sign that relations are improving and will continue to improve.

When Soviets negotiate, they envision a different process. One side's opening statement defines issues on which no compromise is possible. They expect the other side to make a similar statement. A Soviet negotiator will look closely at what has not been ruled out. He will then state another issue on which compromise is not possible. He expects his opponent to do likewise. This process continues until a situation has been defined that neither side has ruled out. This becomes the de facto agreement. The feelings at the end of a Soviet negotiation are completely unlike those at the end of an American negotiation. Each side can feel victorious since it never compromised. At the end of a Soviet negotiation there is a feeling of exhaustion mixed perhaps with exhilaration from a long and hard-fought struggle which had a satisfactory outcome. Rather than a feeling of friendliness among negotiators at the end of a negotiation, the most that can be achieved in a Soviet negotiation is grudging respect for a skilled and committed adversary.

Diplomats who have long experience in dealing with the Soviet Union are familiar with these differences in negotiating styles. But the majority of the American people are not. They assume that the Soviets are like us. They wonder why the United States does not have better relations with the Soviet Union. Meanwhile, each new American administration presents the opportunity for a new group of leaders to experience the frustration of dealing with the Soviets and the anger that follows. The Soviets interpret this anger as hostility and feel justified in continuing their arms buildup.

Because the cultures of the United States and the Soviet Union are so different, people in each country often have difficulty imagining how the social and political system of the other country works at all. Americans believe that centralized planning, the lack of a market mechanism to set prices, and the suppression of the free flow of information would create a system so inefficient it would barely be able to produce anything. Russians believe that a system that permits as much freedom as occurs in the U.S. would be chaotic, unmanageable and characterized by uncontrolled violence.

In a society where high ethical status is associated with a willingness to compromise, there is little need for decisions to be imposed by a central government. However, in a society where high ethical status is associated with refusal to compromise and a willingness to be in conflict with an opponent, a central government with considerable authority is necessary to maintain order. Understanding a system of government requires understanding the culture in which that government is rooted. Similarly, trying to understand how a government may change is not possible unless one has some idea of how cultures change.

Lefebvre's work on the ethical systems in the West and the Soviet Union explains how each system evaluates ethical status and leads to feelings of guilt and suffering, but it does not contain an explanation of how ethical systems change. Kohlberg's theory of moral development does contain an explanation of how people move from one level of moral development to another, namely by encountering problems that cannot be successfully solved at the current level of moral development but can be solved at the next higher level of moral development. Efforts to combine these two theories might be a fruitful line of inquiry for those attempting to understand change in Eastern or Western social systems, for example change from one style of management to another style of management.

9 THE NEED FOR KNOWLEDGE TO MANAGE CULTURAL INTERACTION

Coexistence between the United States and the Soviet Union requires that joint decisions be made. The two political and economic systems are very different in both organizational structure and practices. The different institutions have their foundations in different value systems. There is ample evidence that, when people from the United States and the Soviet Union interact, puzzling and surprising problems arise due to the different value systems.

The result is that interactions which are on-going tend to be superficial. Efforts to understand the differences between the two systems often result in anger, hostility, and the conviction that further interaction would be fruitless. It is time that we attempt to understand why these feelings occur, despite sincere efforts on both sides, and begin to formulate strategies for moving beyond these feelings.

Peaceful coexistence in the long term will require some important changes in values and social practices. In order to achieve this change in a manner that is not disruptive it will be necessary to explicate the value differences. More successful management of the interaction between the two systems will require more thorough understanding. The fact of differences can hardly be questioned. Why do Americans behave as they do? And why do Soviets behave as they do? Though it may be unpleasant for both sides, an explication is indispensable. New forms of conduct need to be established based on new values.

There are indications that there is a willingness on both sides to engage in new conduct. The research described above can contribute to the "new global political thinking" which has been called for in the Soviet Union. Gorbachev has declared that there is a need to give up orthodox approaches of regionalism and narrow class interests. Currently experiments are underway to increase emigration from the Soviet Union, to permit joint ventures between East and West not administered by the government, and even to allow private ownership of some services such as restaurants with owners allowed to keep profits. If interaction between the two societies increases in the years ahead, the success of joint projects will require deeper understanding among a larger number of people than currently exists.

My experience has been that there are major conceptual difficulties in understanding relations between the United States and the Soviet Union. My conclusion is that the focus of attention should be not on the arms race but rather on the cultural and ethical systems underlying the two societies and that at least a few theories already exist to help us formulate fruitful methods and lines of inquiry.

ACKNOWLEDGMENT

This paper benefited from helpful suggestions made by Edward Manukian.

REFERENCES

1. Arbib, Michael and Ernest Manes. *Arrows, Structors, and Functors: The Categorical Imperative.* Academic Press, 1975.
2. Axelrod, Robert. *The Evolution of Cooperation.* Basic Books, 1984.
3. Kohlberg, Lawrence. *Essays on Moral Development: The Philosophy of Moral Development.* Volume 1, Harper and Row, 1981.
4. Lefebvre, Vladimir. *Algebra of Conscience: A Comparative Analysis of Western and Soviet Ethical Systems.* Reidel, 1982.
5. Miller, James G. *Living Systems*, McGraw-Hill, 1978.
6. Mitroff, Ian I. "The Complete and Utter Failure of Traditional Thinking in Comprehending the Nuclear Predicament: Why It's Impossible to Formulate a Paradox-Free Theory of Nuclear Policy," in this volume.
7. Nagel, Ernest and James R. Newman. *Gödel's Proof.* New York University Press, 1958.
8. Possony, Stefan T. and J. E. Pournelle. *The Strategy of Technology: Winning the Decisive War.* Dunellen, 1970.
9. Russell, Bertrand. *Principles of Mathematics.* W. W. Norton and Company, first published in 1903.
10. Umpleby, Stuart. "Some Implications of Cybernetics for Theories of Social Systems," in George E. Lasker (ed.) *Applied Systems and Cybernetics*, Pergamon Press, 1981.

DISCUSSION OF UMPLEBY'S PRESENTATION

Zeleny

If the problem of East-West relations lies in cultural differences, then there is little hope that there will be an improvement in relations in the near future.

Umpleby

Yes, a long-term strategy does seem to be required. But societies and cultures do change. They change in a direction and at a rate that people in the society feel is appropriate. If the origin of the conflict is cultural differences, non-cultural strategies are not likely to be successful.

Zeleny

We tend to delegate questions of conflict to the realm of the psyche (perceptions, values, preferences, world-views). That implies "resolving" conflicts by attacking these psychical "sources of conflict" directly by persuasion, negotiation, re-education, propaganda and brainwashing. No stable, long-term changes in human values and world views can result.

A more natural and more manageable source of conflict could be found in the paucity of decision alternatives we are considering. The point is to invent or design *new* alternatives which would *remove* the conflict while *preserving* the value preferential identity of the agents involved. One has to evolve new alternatives, not "change" cultures to accept the old ones. That has been tried.

Umpleby

I agree. Propaganda based on existing points of view is not helpful. What I am suggesting is that a deeper understanding is required in order to build a foundation for new alternatives. I believe that more attention should be devoted to the relationship between culture and decision processes. For example, how do cultural differences between the U.S. and the USSR lead to conflict while cultural similarities channel that conflict into military competition? The amount of money required to do research and experimentation on the cultural foundations of decision processes would be very small relative to the amount of money we are spending on weapons. I believe a small amount of money spent on collaborative research could have important long-term results.

Rapoport

The idea of paradox will not help decision-makers. They do not have time to think about theoretical ideas.

Umpleby

Officials and managers need time to reflect. In the U.S. the leading officials in the party out of power spend time at universities and at think tanks such as the Brookings Institution. During their time out of power these former and future government officials have time to examine what worked and what did not and to think through new courses of action. Some industrial firms in the West are also beginning to give their managers sabbaticals.

Mitroff

Why only those levels? Do you have a "theory" of levels?

Umpleby

I do not have a well developed theory of levels, but here are a few starting points for constructing such a theory.

1) The levels I am interested in are mental models, some of which are, at least to some degree, subsumed by others. I am not talking about material levels, for example cell, organ, organism, group, organization, nation.

2) As Russell and Whitehead pointed out near the turn of the century, formulating sets which contain themselves generates paradox and undecidability. Consider the case of the social scientist who is constructing theories of social systems. Are his theories his best judgment of how the social system is operating, or are his theories designed to advance his career within the social system? The mixture of motives is formally undecidable to an external observer and probably not entirely clear to the scientist himself. Analyses which take into account the observer and his interaction with the social system of which he is a part are probably the most challenging to puzzle out.

3) Levels of analysis will usually parallel organizational and role distinctions in a society. That is, organizations and professions are different because they use different mental models. Conflict between organizations and people with different roles occurs in part because the mental models they use are incompatible.

Von Glasersfeld

I agree with the distinction of the two "ethics" (compromise within, confrontation without vs. conflict within, comprise without). But I do not agree with the notion that these are distributed according to "cultures", or even in single individuals. Don't you think that what varies with cultures are the areas (topics) where one is applied rather than the other?

Umpleby

Yes and no. I suspect that most people are capable of reasoning using both ethical systems and use both ways of thinking in different situations. However, there are numerous statistically valid studies that reveal cultural differences and some studies which reveal the difference cited in my paper (see Lefebvre's book). Lefebvre's view of cultural differences is a generalization. The dominant mode of reasoning in many individuals in both societies may not fit the generalization. Furthermore, Lefebvre's theory captures only one facet of two quite complicated cultures. Nevertheless, his point of view is supported by survey research and analyses of literature, and agrees with my personal experiences. I would welcome additional research to test the points you raise.

CHAPTER

5

DEVELOPMENT OF CYBERNETICS AND THE THEORY OF COGNITION

Igor A. Apokin

Certain scientific disciplines, from the theory of cognition point of view, may be interpreted, on the one hand, as a supplement of general epistemological principles applied to a concrete situation and, on the other hand, as the application of new scientific results for disclosing the substance and enrichment of other postulates in the theory of cognition. As regards some very important moments of gnosiological examinations there is also the question of the maximum potential and limitations of cognition within the framework of the corresponding science. In the present work, consideration is given to the development of cybernetics in those disciplines with research which presents gnosiologically interesting problems, including the previous works of the author [1-3]. Regarding cybernetics, the author proceeds from the definition of this area of knowledge, increasingly used (though not popularly accepted) in the Soviet school of cybernetics which is to the credit of V. M. Glushkov: "Cybernetics is a science dealing with the general laws of information acquisition, storage, transformation and processing in complex control systems" [4, p. 440].

For a more precise understanding of the subject under discussion, some words about the structure of cybernetics are offered. As pointed out in work [3], taking into consideration the prevailing condition at the beginning of the 80's, the structure of cybernetics, at first glance, may be presented in the form of a three-level scheme, at each level of which there exists a difference in the level of abstractness of the problems. The upper level comprises the mathematical apparatus of cybernetics (the theory of algorithms, mathematical programming, game theory) and its methodological and philosophical problems. The middle level consists of branches dedicated to researching basic classes of control systems (technical, economic, biological, etc.). Engineering cybernetics occupies a leading position among such branches, including a complex when considered fully or partly developed. This level also includes such theories as the theory of automatic control, information transfer theory, finite automata theory, pattern recognition theory, theories of artificial intelligence and others. The lower level consists of disciplines which study concrete systems (computer science, theoretical robotics and others).

Regarding a theory of cognition, the author operates from the fundamental principle of dialectical materialism (as a science of the most general laws of nature, science and thinking development) about the coincidence of dialectics, logic and the theory of cognition. This thesis is based on the fact that human thinking and the objective world about us are geared toward the same laws and consequently cannot contradict each other. As a result, categories of dialectical materialism, on the one hand, have ontological substance and, on the other hand, have an epistemological function which turns out to be the method of real cognition, reflecting also (analyzing them in a definite sequence) the historical and logical phases of the cognitive process.

1 CONCRETIZATION AND ENRICHMENT OF THE CONTENT OF PHILOSOPHICAL CATEGORIES

Considering the effect that cybernetics has on the exposure of substance, enrichment and development of philosophical categories, it is expedient to approach this question from the analysis of cybernetic methodological function, which is in turn deduced from its examination as regards the leading cybernetic principles and methods.

The fundamental principles of cybernetics can be correlated to the ideas of general feedback mechanisms in different types of control systems and to information processing as the basis of control processes, including the working out of solutions and their realization. Without the necessary material base in the form of cybernetic technology, presented in the form of computers and control systems, the fundamental ideas of cybernetics would not have had great influence. This, particularly, is clearly seen in the leading method of cybernetics—computer simulation—the effectiveness of which is defined in terms of the content of computer programs. The method of computer simulation is linked with other methods which are formulated within the confines of cybernetics, including the method of algorithmization and that of the blackbox and others.

Operating from cybernetic principles and methods, it is possible to approach the question of the methodological role of cybernetics. Its deep analysis is presented in work [5]. At this juncture, too, it is appropriate to say that cybernetics, in the first place, brought a recognizable change into the general scientific picture of the world (partly because information is presented as an important component of such a picture). Secondly, it fulfills the integrative function relative to other scientific disciplines. Thirdly, its methods warrant humanization of science as a whole.

The development of science as a whole charts two routes: "the Babylonian", orientation toward the inductive method and the differentiation of knowledge at the expense of discovering new phenomena and facts; and "the Greek" orientation toward the hypothetico-deductive method and the establishment of links between different factors, theories and branches of knowledge. Cybernetics serves as an important impulse in making science toward the Greek course. Its principles as well as its methods embrace objects and processes of different nature.

We may emphasize that cybernetics serves as a theoretical basis of a very large class of technical means (control systems). According to Ashby, cybernetics relates to real machines as geometry does to real objects [6].

One of the main integrating strategies of cybernetics is the mathematization of knowledge, especially with the aid of computer modeling. It may seem to be a paradox, but the development of mathematization based on cybernetic methods sometimes contributes to the humanization of science, making exact knowledge closer to the humanities. Fuzzy algorithms enrich the algorithmization method by introducing fuzzy notions typical of the humanities and everyday thinking. Fuzzy algorithms make it possible to model many processes which are impossible to analyze with the aid of the usual deterministic algorithms. In general the modeling of thinking has humanized logical instruments of modeling. This trend is traced from 1956 when John von Neumann introduced the time-factor into logic [7].

So the methodological aspect of cybernetics has much in common with general systems theory, as it was formulated by Dr. Sadovsky. I quote: "This is the intrascientific methodological reflection" [8]. The methodological aspects of cybernetics and philosophy differ. Philosophy deals with general principles of human activity, and cybernetics concerns the concrete methods of science in general. This difference makes it possible to consider cybernetic principles, laws and categories as well as their enrichment. Below are some examples.

As a result of the comprehension of the deep relation of information with the reflection principle which constitutes the basis of the materialistic gnosiology, we have the possibility of concretizing the philosophical conception of the ideal. The nature of it, according to Norbert Wiener, is the difference between the signal and its content, that is the information [9]. On the pre-psychical level, it is senseless to divide them—the signal and the information. But on the level of human consciousness there is a separation of the signal and the information [10]. The concrete scientific conception of the ideal consists in information analysis within the framework of some theories, for instance semantic and pragmatic ones.

The creation of "the model" as a scientific conception (in the course of the development of computer simulation) enriched the scientific interpretation of the conception of philosophical laws. Earlier this conception was limited within the framework of scientific laws. The models differ to a great extent from the scientific laws which are "either true or untrue, and it is senseless to discuss whether the laws are "good or bad" [10, p. 356]. The model may consider many more events than a single law or its complex.

The materialistic interpretation of causality is based on matter and the transfer of motion from cause to effect. As a result, the effect influences, to a certain extent, the cause. In cybernetics this influence is concretized by the notion of feedback. The theory of feedback systems makes it possible to express complex forms of causes and effects as well as their interactions in mathematical and scientific language.

The examples show the richness of epistemological conclusions made on the basis of cybernetics research. But one of my ideas is that the evaluation of cybernetics and its philosophical importance from the present positions does not allow us to solve the problem of the maximum potential for cognition aided by cybernetic methods. That is why I think we should try to analyze its most probable prospects.

2 ON THE MAINSTREAM OF CYBERNETICS AND THE MAXIMUM POTENTIAL FOR COGNITION BY CYBERNETIC WAYS AND MEANS

One of the main cybernetic epistemological problems is that of the maximum potential and limits of cognition by cybernetic theories and methods. The close relation of cybernetics and practice helps to solve the problem. Cybernetic practice is the criterion of truth of cybernetic ideas which is expressed first of all in the creation of technical systems. But as distinct from physics, chemistry, biology and other classical scientific disciplines, cybernetics is a very young scientific discipline, and having received only the first results, it is very difficult to make a concrete conclusion about the final ones.

What should be considered the mainstream of the development of cybernetics? To my mind, without any doubt that is the problem of artificial intelligence (AI). The practical solution of this problem is now connected with fifth generation computers and third generation robotics research. Why should we consider this problem to be the most important? We know that the enthusiasm in the development of cybernetics, all the innermost ideas of its creators (creators like Norbert Wiener, John von Neumann, Alan Turing, W. Ross Ashby and others), were concentrated around one idea: to create AI systems more efficient than the biological ones as regards their information abilities. In essence it was a challenge which covered tens and hundreds of millions of years of evolution. The deep philosophical sense of cybernetic ideas was reflected in the very fact of the scale of intention.

Since a generally adopted definition of AI is absent, I propose the following: intelligence comparable with a human one in its abilities, created by means of a certain technology.

I would like to add that the comparison of the abilities of artificial and human intelligences could be understood in two different ways. One way would be a general comparison, that is we create a certain equivalent of natural (human) intelligence, a certain universal AI, either by means of another technological method or we do it on other technological (non-biological) basis. This is the first understanding.

The second way is to compare different functions. In other words we create many varieties of AI, each of them solving only one type of intellectual problem, but much better than the human being. This method involves creating specialized varieties of AI.

If we investigate the problem of the universal non-biological AI, considering its creation from the epistemological point of view, we may conclude that there are no real philosophical arguments denying the possibility of creating such AI in principle. On the other hand there are no arguments proving such intelligence could be created. In other words, it is a scientific and social problem but not a philosophical one.

Let us remember, for instance, what the great Archimedes said about the ability to move the Earth by means of a lever. This problem is unsolvable from the epistemological standpoint (there are no essential philosophical objections; however, neither are there arguments justifying the possibility of a practical solution). The problem belongs to the domain of science and technology and may be theoretically solved, for example, by way of mathematical simulation. Also, its social dimension should be born in mind. It is expedient to speak about practical solutions only in cases when they are of social value, not merely of abstractly cognitive value. A certain analogy, although incomplete since we are dealing with far more complex phenomena and their consequences, arises also regarding AI.

I have my own view on the means for creating AI which is connected with epistemological problems.

In a brief account, my position assumes that the experience of the formation of natural intelligence during millions of years is too important to be ignored when creating artificial intelligence. This experience suggests that the individual's intellect is formed, first, in the social environment, and second, in the processes of learning and self-learning. These processes are necessarily based on sensual experience which is indispensable for the formation of the seemingly elementary notions of the surrounding objects and processes ("water", "stone", "to go", "to break", etc.) of three-dimensional space—the essence of the simplest arithmetical operations (like 1 + 1 or 1 : 2)—etc. Further, the processes of learning and self-learning are inseparably connected with the emotional and motivational spheres. Thus, it is difficult to expect any success without an elementary inquisitiveness. Finally, the processes of learning and self-learning are valid only in relation to the material substratum (brain) of a certain level of complexity and structure. Nothing can be done beyond some complexity threshold of the brain. This was demonstrated by experiments when a baby monkey and a human baby were brought up from birth, and educated together.

In application to AI and taking into account the modern way of its realization on the basis of the solid body, this sketch of reflection suggests that a universal AI should have the following features:

1. A structure should have at least three subsystems: control, sensor, and executive. With regard for the structural asymmetry of the brain, it appears that two units should form the basis of the structure of the control subsystem: a "computing" unit, that resembles (practice will show to what extent) a complex microprocessor system of the superlarge integrated circuits, and an "associative" unit which might be realized on

the basis of holographic technology. It is evident that the level of sensors should correspond to the potentialities of the computing and associative units. As regards the executive subsystem, it seems impossible to do without mechanical units providing advanced manipulating potentialities and the mobility of the system as a whole.

2. It is difficult to define *a priori* the complexity threshold of the computing unit. One might only assume that it is to be found somewhere on the level of 10 billion bytes (i.e. about 5 million pages of text) for the operational memory capacity and one trillion operations per second, secured, in particular, by a more complicated structure.

3. Built-in algorithms, determining a hierarchical system of the most general goals (for example the primary orientation towards the search for the solution of problems, the study of the environment, self-learning, etc.), as well as the ways of their achievement. In essence, these built-in algorithms determine the AI axiology, the hierarchical system of values, that will necessarily be present.

4. The initial stages of learning and self-learning ("training") are conducted within some specially arranged environment, which might also include other analogical or related AI systems. It is clear that the optimum character of this environment cannot be established a priori.

Creation of such AI would be obviously a revolutionary step towards the development of science and technology. The importance of achieving this would lie not only in the fact that we would create an AI compatible with the human intellect, but also the created AI, as any other technological system, would be capable of a more or less rapid improvement, an increase in "intellectual power". However, before proceeding to the philosophical comprehension of such developments, let us say a few words about whether this event is feasible.

As stated above, the possibility of creating AI is solved not with philosophical arguments, but through scientific studies and on the basis of social needs. Treating the problem as a scientific one cannot fail to reveal that it is solvable in principle, since "any process including that executed by our intellect, of which we are fully aware by the way this process proceeds, may be simulated in principle on the discrete automaton" [11]. However, this statement of the question by no means implies practical realization (just as theory does not prohibit building a tunnel along the Moon's diameter).

What can be said about practical possibilities? An optimistic view (the one shared by the author) is based mainly on the actual course of events. During the last 30 or 40 years (that is in a very short period on the scale of natural evolution) it became possible to build computer systems of enormously high speed (about one billion operations per second) on a solid technological basis (in laboratory specimens of semiconductor integrated circuits the arrangement density reached a million transistors per square centimeter). These milestones, which quite recently were regarded as unattainable, became reality. Of course, we are still far from obtaining the characteristics of neuron ensembles in the cortex, but "the light at the end of the tunnel" is already in sight. A thousand fold improvement of some indices of integrated circuits (mainly as regards the arrangement density and scattering power), as well as a substantial complication of the structure of the links between the circuit elements, will provide us with the material necessary for overcoming the "complexity threshold" mentioned above. Technological difficulties are still innumerable, but the rapid rate of overcoming them does inspire hope of success.

Of course, our discussion concerns only one, technological aspect of the problem, but its successful solution will serve as an essential basis for further work. It is evidently far from the basic to the whole edifice. Besides, not only possibilities are important but also the desire to realize them. It is on this point that we pass from technological and scientific problems to philosophical ones, dealing with the place of man in the world, his future and the ultimate potentialities of his knowledge.

It is expedient here to approach the hypothesis discussed in [3]. Its main point is that the possibilities of a more and more radical influence on man will arise with the progress in science and technology, while some of these possibilities will be realized with humane purposes (treatment of hereditary diseases, replacement of disabled organs, extension of human life, etc.). In the long run this will result in the man of the future being substantially different from the man of today. In other words, the main premise implicitly included in the majority of considerations of the future is false. That is the premise that while changing the world to an ever growing extent, further penetrating into space, and technologically mastering more and more powerful forces of nature, man himself will remain always the same. Just as in the previous biological evolution, a process of transformation will take place. However, contrary to evolution, which is indifferent to value systems other than survival (categories of happiness and the good), this process will proceed consciously, being subordinated to the collective mind of the future civilization and oriented towards man as a goal, not as the means of achieving other goals. On this point the discussed hypothesis fundamentally differs from other ideas regarding technology such as the notorious concept of the cyborg (cybernetic organism).

In principle, there exist two methods of transformation: biological and technological. Improvement of biological methods might result in a civilization which makes it possible to solve many human problems within its own framework: diseases, premature aging, etc. By improving the genetic apparatus and developing methods of upbringing and education within the framework of such a civilization, a substantially higher average level of abilities may be achieved. Possibly, a considerable extension of the average length of life may be attained by research along these lines.

A growing and primary application of the technological methods for purposeful influence on man is an alternative method of human development which might result in a technological civilization with man changing himself consciously and gradually. The motives of such changes may be numerous: aspiration for extension of life (if biological methods fail here) or for direct (not only by means of automata) exploration of nature in extreme conditions (space, ocean) etc. It is not necessary to give a more detailed list of stimili and versions of technological changes, since they are not very predictable today and rather belong to science fiction. Here it is important to emphasize that the tendency toward technological changes does exist and is dictated today by the natural and humane goals to create artificial organs instead of disabled persons.

Outlines of the new revolution, even more radical in its consequences, are coming into view behind the horizons of the modern revolution in science and technology. It is man himself who is an object of this revolution, which is humane in its essence, while its goals are guarantees of health and life extension, improvement of physical and spiritual potentialities, and the development of creative abilities. Cybernetics plays here a dual role. First, biological cybernetics, medical cybernetics, and social cybernetics will be more widely used for the solution of specific problems. Second, more and more complicated AI systems created by technological cybernetics might provide us with the key to some very radical forms of technological changes of man that are consonant with human hopes and aspirations, as well as create a certain psychological atmosphere in which such transformations would be viewed as a new stage in the development of civilization.

The range of epistemological conclusions that may be drawn on the basis of the presented hypothesis is obviously extremely wide. The main conclusion is that if the development goes in the discussed direction, the ultimate potentialities of cognition by cybernetical means exceed the horizon of the visible so that at a certain stage they link up with the ultimate potentialities of human cognition.

However, even if one does not proceed from the proposed reasoning but confines oneself to the superficial assumption which is near to us in the time of realization (less than half a century), that is the idea of creating a sufficiently powerful universal AI, the scale of epistemological consequences would be still rather vast. First, the notion of the potentialities of human cognition becomes wider. Second, with the development of AI, the issue of "computer cognition"—its specific features, potentialities and limitations, and its relationship with human cognition—will become one of the central issues in epistemology.

Although AI is built by man "from beginning to end", in the course of its development it will obtain even newer features with their analysis providing conclusions of epistemological significance. Irrespective of our will, powerful AI systems will present mankind with problems which will make us come back to the discussed line in the relationship between human development and progress in science and technology.

REFERENCES

1. Apokin, I. A., Scientific and Technological Progress and Prospects of Biological and Technological Evolution of Human Beings, in *The Scientific and Technological Revolution. The Prospects*, pp. 130-138, USSR Acad. Sci., Moscow, 1980.
2. Apokin, I. A. The Present-day Scientific and Technological Revolution, Its Contents and Prospects, *Priroda (Nature)*, pp. 66-73, No. 4, 1981.
3. Apokin, I. A., *Cybernetics and Scientific and Technological Progress (history and prospects)*, pp. 1-248, Nauka Publishers, Moscow, 1983.
4. Glushkov, V. M., Cybernetics, in *Encyclopaedia of Cybernetics*, pp. 440-445, Kiev, v. 1, 1975.
5. Biriukov, B. V., *Cybernetics and Methodology of Science*, pp. 1-414, Nauka Publishers, Moscow, 1974.
6. Ashby, W. R., *An Introduction to Cybernetics*, pp. 1-432, Foreign Literature Publishers, Moscow, 1959.

7. Neumann, J. von., Probability Logic and Synthesis of Reliable Organisms Out of Unreliable Components, in *Automata Studies*, pp. 68-139, Foreign Literature Publishers, Moscow, 1956.
8. Sadovsky, N. V., Problems of Systems Methodology in Contemporary American Philosophy of Science, in *Systems Research*, pp. 84-108, Nauka Publishers, Moscow, 1981.
9. Wiener, N., *Cybernetics*, pp. 1-324, Sovetskoe Radio Publishers, Moscow, 1968.
10. *Control, Information, Intelligence*, pp. 1-383, Mysl Publishers, Moscow, 1967.
11. Biriukov, B. V., On Abilities of Artificial Intelligence, *Vorposi Filosofii (Philosophical Problems)*, N 3, pp. 88-91, 1979.

DISCUSSION OF APOKIN'S PRESENTATION

Umpleby

In our country in cybernetics we discussed learning in the 1960's, self-reference in the 1970's, and autonomy in the 1980's. Researchers in artificial intelligence in this country are just now making the transition from clever programming to serious consideration of learning. Has there been a similar evolution of thinking in cybernetics in the USSR, and is there a similar lag between cybernetics and artificial intelligence in the USSR?

Apokin

If we speak about general paths to artificial intelligence, these are *almost the same* in the USSR and in the U.S. Three main paths have been: 1) transistor models of neuron ensembles, 2) computer programs, 3) simulations of evolution. These three approaches to AI are investigated both in the U.S. and in the USSR. For details concerning theoretical principles and real achievements it may be better to listen to the specialist on AI—Dmitry Pospelov.

Pospelov

There has been a coevolution of implementation and theory. In Soviet AI research there have always been two types of specialists—pragmatists and theorists. The pragmatists felt, "We must do what has to be done, any way we can." The second held, "We must do what we can, and the way to go is to understand how the human brain operates." At the first stage of AI development (1965-1975) the pragmatists were more influential. Programs were set up for solving concrete problems illustrating human activities. But these were not sufficiently general AI models. After 1975 the situation changed. AI systems started to be created for resolving whole classes of tasks. This required elaboration of models of intellectual activity. Now in the USSR we are actively elaborating such models. Cyberneticians, psychologists, and linguists are working together.

We have a joke that cyberneticians make more precise mistakes than others.

DISCUSSION OF EPISTEMOLOGY

Apokin

Second order cybernetics is more general than cybernetics. It concerns science as a whole.

Allyn

I would like to relate several points made thus far, particularly the notion of paradox articulated by Mitroff and the discussion of "levels of thinking" introduced by Umpleby. The latter stressed the idea that a problem often must be resolved on several levels. If this approach is taken, certain of the "paradoxes" noted by Mitroff cease to be so.

To put this in a specific context, let us examine the "paradox" that greater capability for verification of weapons character and location decreases security given the increasing accuracy of ICBMs. This very argument against increased verification is currently being put forth by the Soviet side at the Stockholm Talks on Confidence-Building Measures. Currently, the U.S. is proposing several measures to increase the "transparency" of both sides' military forces. The Soviet side objects to these measures in the context of the current poor U.S.-Soviet relationship, suggesting that such measures would enable military staffs to choose targets with greater accuracy. If the U.S.-Soviet relationship were to get better, however, the Soviet side asserts, then these same verification measures would not have a negative effect but would *increase* security. Thus the "paradox" dissolves if the problem is approached not solely on the level of military thinking, but on the level of political thinking (action to improve the U.S.-Soviet relationship).

As Rapoport suggested, these "paradoxes" may be a result of looking at policies in a limited context, and when approached from several levels of analysis, they cease to be paradoxes. Also, I think that we should define with more rigor what we mean by "paradox"—is it a logical contradiction, or an action which produces a counter-intuitive result, or something else like the paradox, "in order to keep the peace, prepare for war?"

Mitroff

The context has to be given before resolution is possible. New paradoxes will emerge as we move to higher levels of analysis. Perhaps they will be more manageable. I want to push the position.

Reckmeyer

The problems are multidimensional. They require a respect for multiple perspectives.

Von Glasersfeld

Is it possible to make a computer reflect on its own operation?

Pospelov

Ego is possible only if there are two independent systems of perception. In an individual there is perception of what is outside and symbolic processing. We can use words to represent emotions. Only on this condition can the system sense its own ego. Then it can conceptualize about itself. Today, AI models the processing of symbols. We do not now know how to pose the problem.

Lapin

The first day's discussion has demonstrated the ability to discuss complex and new questions at the international level with participation by specialists in systems methodology, cybernetics, philosophy, etc. We should give thought to how to continue discussions of such problems by *scientists* interested in identifying these complex, even paradoxical, issues and in seeking new and more effective methods of solution. This presupposes a search for new forms of communication between scientists—both within countries and at the international level (especially between U.S. and USSR scientists). I have in mind, in particular, non-traditional heuristic forms of communication. I will discuss this in more detail in my paper about innovation.

Munipov

The many different problems we have been discussing here can be reduced to two classes. One is that if human progress is considered relative to technical progress, man's progress to date would have to be considered zero. The other is that technical progress can lead to a situation in which war is unthinkable. However, this does not mean that we should not give thought to how to prevent a nuclear catastrophe. Furthermore, we should not delay in taking cardinal measures to make war impossible. The systems approach and modeling will help us to a better and deeper understanding of these problems.

Sadovsky

Several years ago I wrote an article on paradoxes of system thinking. It was criticized. Some paradoxes were said not to be paradoxical.

The term paradox has been used in many ways in our discussion. It is very difficult to come up with only one meaning. We need to investigate different kinds of paradox. This could be a subject at the next meeting which could be devoted to urgent world problems. There is a good basis for conversations.

Second order cybernetics has introduced a great change in cybernetics and general systems theory.

We should distinguish two kinds of scientific activity. 1) Observing with cybernetics and systems techniques. 2) Developing knowledge and tools.

Mitroff

It is not necessarily true that when we go to a higher level of analysis the paradoxes will be resolved. We shall find all of the same kinds of paradoxes when we come to global problems. Can we teach leaders that by taking a broader view they can cope with complexity? Our systems constrain us from confronting paradox. The purpose of systems theory was to confront complexity. We cannot defer any longer the metaphysical problems.

Reckmeyer

One qualification on the shift to a higher level—mixing the immediate and the not-so-immediate.

People need to become comfortable with shifting levels of analysis.

The systems approach gives us a way of talking to each other.

Ben-Eli

A problem is not something shameful. People need to learn to live with them. When people do not admit they have problems, things get worse.

Mitroff

If we want to guard products from psychopaths, we need to bring in psychopaths to help us design packaging to block psychopaths. I am like a doctor for companies. If they cannot face a situation, they cannot deal with it. We do not teach people how to handle intellectual or emotional complexity.

Reckmeyer

Traditional science is answers oriented. The systems approach is questions oriented and heuristically oriented.

Von Glasersfeld

Regarding constructivist and realist conceptions of knowledge, in our discussion "problems" float about like crystals. But paradox is not there "in the world." It comes from our categorizations and our goals. It can be cured by deciding where we want to go.

Sadovsky

The only way in mathematics is to avoid paradox. There is no great difference between constructivist and realist points of view in this connection.

Von Glasersfeld

Some paradoxes in logic were resolved by changing the rules, for instance by introducing the notion of time.

Umpleby

Bertrand Russell developed his idea that self-reference leads to paradox when he was writing *Principia Mathematica*, which uses set theory. A more recent mathematical approach, category theory, uses arrows and hence contains an implied time dimension. Assuming that an entity can fit one category at one time and another category at another time, can be one way of resolving paradox.

Mitroff

The stress here has been on the problem formulation side. American culture breeds certainty addicts. We should pursue questions. We have incredible tunnel vision.

Lapin

It is true that the systems approach enables us to see things in a new light. But we cannot limit ourselves to modeling. We must also support those who want to solve problems. Sometimes we do not have the methods we need. New methods need to be developed and old methods elaborated.

Rapoport

Science is not able to resolve all problems yet. There is still work to be done.

CHAPTER

6

ON SOME NEW LINES OF RESEARCH IN CYBERNETICS AND SYSTEM MODELING

Oleg A. Kossov

The purpose of this presentation is to outline the major lines of research activities conducted in one of the Institute's departments.

First on the list is cybernetics, or more specifically, feedback systems methodology. Figure 1 shows a classical structure of a closed-loop control system with a negative coordinate feedback where plant L_o is controlled by a signal $u(t)$ which results from the dynamic transformation L_u of the error signal $x(t)$ which is the difference between the control input $g(t)$ and the output $y(t)$

$$u(t) = L_u\ [k, x(t)]\ , \qquad (1)$$

where k are parameters of the operator L_u. And for the general case,

$$x(t) = \psi\ [g(t), f(t), a(t)]\ . \qquad (2)$$

If the plant parameters are constant or variable, but known, then the impact of the uncontrollable exogenous forces $f(t)$ belonging to a known class of functions $F(t)$, can be reduced to the desired range or eliminated completely by the appropriate selection of the coordinate feedback operator L_u featuring variable parameters. Hence, in

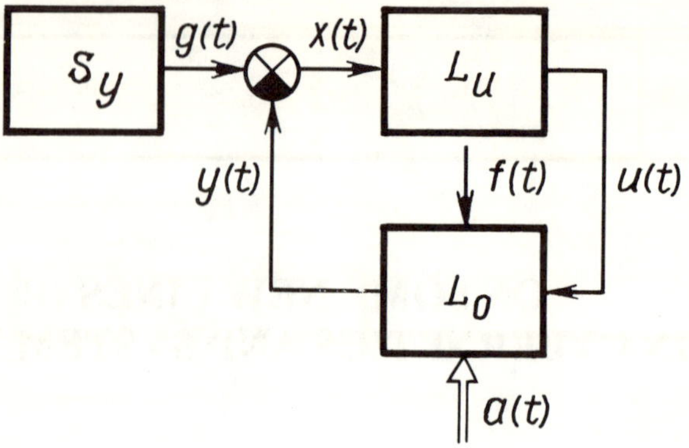

Figure 1

this case the feedback is a tool for eliminating the uncertainty in the plant behavior caused by the unknown coordinate disturbances $f(t)$.

If the plant parameters $a(t)$ vary in time in an unknown way and the ranges of their variation are fairly wide, then the choice of a proper feedback in Fig. 1 is considerably complicated. Under such circumstances, the constant feedback does not do the job and the need may arise to use, for different values of $a(t)$, their own feedback equation (1) differing not only in the values of parameters k but possibly in the form of L_u conversion.

One of the ways to eliminate the impact of parametric disturbances of $a(t)$ and coordinate disturbances of $f(t)$ on the control process is to use, as applied to the structural scheme of Fig. 1, an intensive negative feedback. The desired result is accomplished, within the framework of this approach, in the limit when the feedback gain in (1) is infinite. In real life, such closed-loop control systems are as a rule unstable for two reasons:

- the impact of small unrecognized parameters happens to be significant;
- nonlinearities emerge as a result of physical constraints on the elements.

The coordinate feedback (CF) can thus effectively offset only the coordinate disturbances.

The adaptive control methods for nonstationary plants provide for the desired form of transient processes with finite gain in the feedback loop and stability. However, the decisive factor here is the quasistationarity condition which assumes slow variation of the plant parameters whose current values can be measured directly or indirectly in order to generate an efficient control law for a subsequent time interval.

The Institute researchers suggested a new approach to the control of nonstationary plants whereby the feedback principle is consistently used and helps remove constraints on the variations in the plant parameters [8].

The basic idea is presented in Fig. 2. The structure elements S_y, L_u, and L_o make up a conventional closed loop; the data on disturbances of $f(t)$ and $a(t)$ are in short supply or unavailable, while the operator L_u performs the simplest dynamic transformations with a limited gain.

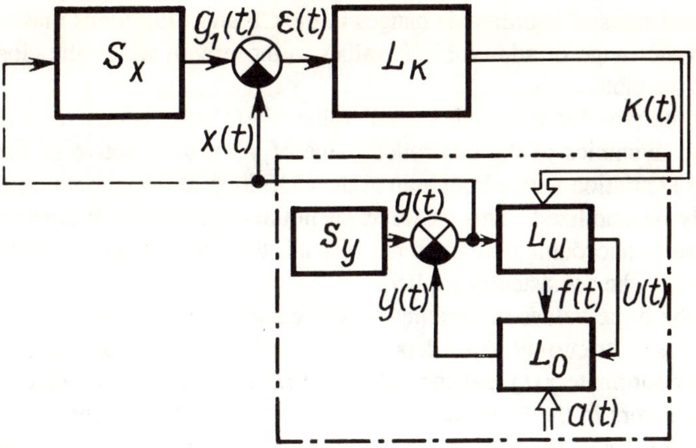

Figure 2.

The desired transient process is obtained at the output of the plant L_o by employing an additional loop—a coordinate-operator feedback (COF) which includes a setpoint unit S_x, an operator L_k, and operator L_u in the main loop.

Let the closed-loop system (1), (2) feature the desired dynamic properties even with variable $a(t)$ if at each instant in time the error coordinate $x(t)$ satisfies the condition

$$x(t) = S_x\,[c, x(t)] \tag{3}$$

where S_x is a dynamic operator whose form and parameters c have to meet the requirements on the control system. If the operator S_x is known, then the setpoint in the COF loop is

$$g_1 = S_x\,[c, x(t)]\,. \tag{4}$$

Then the misalignment

$$e(t) = g_1(t) - x(t) \tag{5}$$

will characterize deviation of the actual behavior of the system (1), (2) from the desired one for $e(t) = 0$. In error-closing control, the error $e(t)$ is compensated by the negative feedback whose outputs are values of parameters k in equation (2), i.e.

$$k(t) = L_k\,[\alpha, e(t)]\,, \tag{6}$$

where α are parameters of the feedback operator L_k which has a variable structure in a general case.

The coordinate-operator feedback has a number of specific features:

- the plant coordinates (error and its derivatives) in the COF loop are the transfer coefficients, they should be of constant sign which calls for changes in the signal signs when the coordinate signs change;

- the plant operator L_o outside the COF loop which ensures its high speed;
- the different rates of coordinate changes in the CF and COF loops make it possible to extend the range of admissible k values to those with which the closed system (1), (2) is unstable.

It is quite possible to secure a small value of $e(t)$ and, hence, meet the requirements on the dynamics of the controlled value of $y(t)$, irrespective of the nature of $f(t)$ and $a(t)$ variation with a high gain in the COF loop as the parameters S_x, L_k, and L_u can easily be stabilized. This, however, is not always feasible because of physical constraints and unaccounted parameters. Instead, the principle of error-closing feedback control may be consistently applied.

With the finite values of parameters α in equation (6), the error $e(t)$ is nonzero in a general case. Its current value depends on two factors: the transient process of the plant output coordinate $y(t)$ and the value of $f(t)$ and $a(t)$ disturbances. Since the setpoint parameters in the COF loop (4) are invariable and the plant parameters $a(t)$ vary over a wide range, the effectiveness of this feedback varies with the plant state.

The component of error $e(t)$ resulting from the plant parameter variations may be reduced by using the operator feedback (OF), as shown in Fig. 3. As $e(t) \to 0$, the values of k in (1) correlate in a certain manner with the plant parameters $a(t)$. Accordingly, by relating c and k through the operator

$$c(t) = L_c\ [\beta, k(t)] \qquad (7)$$

where β are L_o parameters, the desired result can be obtained.

The reduction of the error $e(t)$ by varying the setpoint parameters c makes the system's dynamics dependent on the plant parameters. To keep the dependence in the desired range, parameters c may be constrained. If, however, this dependence is to be minimized or eliminated, yet another type of feedback (Fig. 4) may be employed, where the setpoint unit S_c and the operator L_v ensure an operator-coordinate feedback (OCF) which acts as follows.

Let C_o be some parameters determining the best, in a sense, properties of the closed-loop system when the condition (3) is met. Then the difference $\Delta c = c_o - c(t)$ characterizes deviation from these properties. The misalignment is reduced by subjecting the plant to an additional signal

$$v(t) = L_v\ [\gamma, \Delta c(t)] \qquad (8)$$

where γ are parameters of the feedback operator L_v.

Physically, OCF acts directly on the plant when the setpoint parameters deflect from the best ones in the transient process. In this way the difference $\Delta c = c_o - c(t)$ is reduced or, under floating control, eliminated. The system properties can thus be made fully independent of the disturbance with finite transfer coefficients in the feedback loops.

The considered approach proved rather effective in handling theoretical problems and in developing control systems for metallurgical and petrochemical processes. Article [9] clearly illustrates the potential of this approach in analyzing the problem of aperiodic stability of a nonstationary plant described by the equation

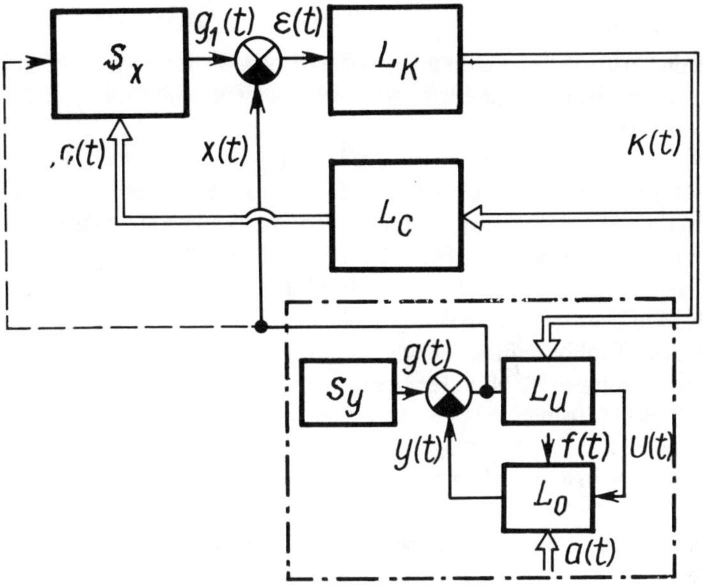

Figure 3.

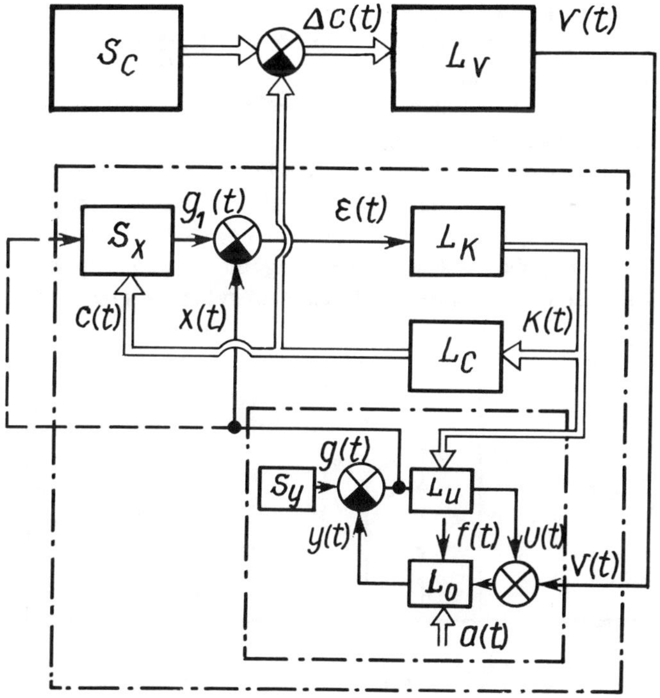

Figure 4.

$$T^2(t)\ddot{y} = u \text{ for } 0 < T_1 \leq T(t) \leq T_2.$$

The second part of the presentation is concerned with new ways to solve a conventional systems analysis problem—the development and choice of preferable plant design options.

As a rule, such a problem is solved in the following sequence: the purpose of the project is determined, the fullest possible set of options is developed, the set of evaluation criteria and a decision rule are assigned, the designs are evaluated and compared so as to select the preferable ones. Sometimes, this is the only possible procedure, but it is extremely labor-intensive if the number of alternatives to be chosen from is large.

In our research we look for such situations of development and choice that require considerably reduced efforts due to, for example, a direct generation of only the assigned number N_o of preferable or admissible alternatives.

We shall illustrate the procedure with an example, where:

- the objects have a morphological structure;
- the elements in morphological classes and objects are evaluated by a common set of criteria;
- the estimates of the object alternatives are functionally related to the estimates of the elements.

Let the object options be made up of elements which are members of L morphological classes, each class having $P_{\min} \leq P \leq P_{\max}$ elements evaluated by n criteria. The criteria are ranked by importance $K^1, K^2, \ldots, K^n$ which generates a lexicographic preference relation on option estimates which are in our case the additive functions of the element estimates.

Let us show that the algorithm generating N_o of preferable and lexicographically ordered alternatives can work.

Rank the elements in each class by the first criterion estimate. The result of this step is shown graphically in Fig. 5 (the case of four classes, three criteria, and three or four elements in each class) where the points connected by lines represent an estimate of an element m_p^l, the superscript denoting the membership in a class, and the subscript—the sequential number of the element in the class ranking.

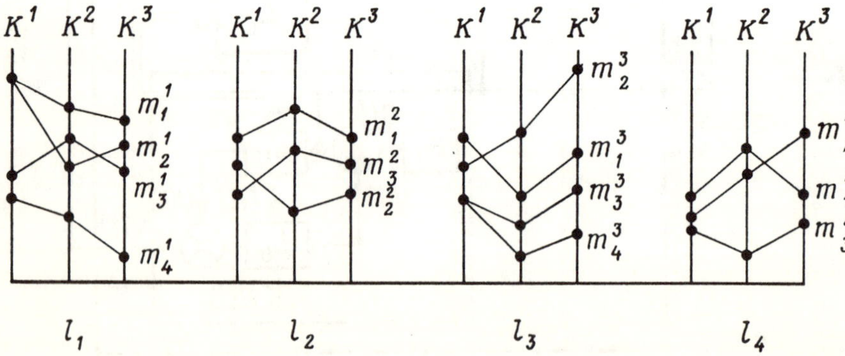

Figure 5.

Generate now a lexicographically maximal option by taking, from every class, an element with the highest first criterion estimate. Determine in every class the difference of estimates by the first criterion, the highest estimate by this criterion in this class being the minuend. Assign to every difference d_p^t an index (the superscript and the subscript) of the element which is the subtrahend. Rank all the differences in increasing order. Combine the least differences into three to five pairs (triples) and assign to them the indices of the two (three) elements in this combination. Enter the composite differences into the last ranking.

In the class of elements with the minimal difference, replace the element in the maximal option with the one whose index was assigned to the minimal difference. Proceed in the same way in replacing the element of the maximal set with that which forms the next least difference. The procedure continues and as early as at the third step a minimal pair of differences may be found which necessitates the replacement of the two elements of the maximal option.

Note that at each step of the procedure (especially with a discrete criteria scale) several options having the same first criterion estimates may be obtained. If the total number of options available at the time is less than N_o then all the options of the current step are included in the list. If this number exceeds N_o then the last group is ranked by the second and, if necessary, the subsequent criteria so as to arrive at the desired number N_o of options. If several maximal options are obtained at the first step, then the subsequent steps are performed with respect to each one of them.

The procedure continues until the desired number N_o of options is obtained or until the elements which make the last pair (triple) of differences included in the common ranking are included in the set. In the latter case, the ranking includes the next two or three least pairs (triples) and the procedure continues.

Consider another example. In the same situation of generation and choice let the criteria be equal and the ranges specified within which the object estimates have to stay (Fig. 6). If the estimates are in this range, the option is regarded as admissible. In this case it is not required to find the "best" options because the purpose is to ensure a balance of estimates by a set of criteria with specified accuracy.

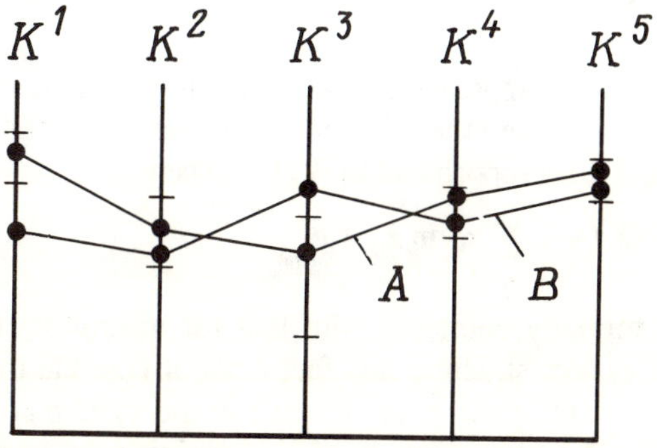

Figure 6.

A heuristic algorithm based on the adequacy principle was suggested for the problem. The idea is that the elements of every class are tested in a certain sequence; their estimates are compared, for example, with minimal requirements to system estimates, and the minimal r.m.s. deviations are determined for all criteria. The element meeting this requirement is the first to be included. Its estimates are subtracted from the minimal system estimates, and the element estimates in the remaining classes are compared with the difference thus obtained. The procedure continues until representatives of each class are included in the system. If the estimates of the resultant option are within the specified range, the option is regarded as admissible (A in Fig. 6) and included in the list, otherwise (B in Fig. 6) it is rejected. The next option is generated from the element which is second by the "similarity" attribute.

Various applied problems were solved by using these approaches. Thus, in generating lists of higher vegetation for a closed cycle life support system designed to meet human demands for all necessary vegetable food, five to ten admissible options of five to eight plants had to be obtained from an original list of 28 to 35 plants. The possible options total 25×10^6 and cannot be handled by search algorithms. The directed generation algorithms turned out to be more efficient [2,4,7]; some of them made use not only of heuristics but also of formal reasoning which extended the known approach on the metrization of morphological space [1].

The third section of the presentation deals with entropy models for the analysis of urban systems development. Proceeding from Wilson's ideas [6] we take account of the multiobjective movements, interaction of entropy subsystems, and combine normative and behavioral modeling.

We introduced the notion of a stochastic communication system [3] generalizing the notion of entropy systems [6]. In this approach a finite set of N elementary particles (e.g. people viewed as labor resources) move from a finite number of volumes of a certain space (origins i) into other volumes, possibly of different space (destinations j). The particles select at random the origin i, destination j, hence the communication ij.

The flow matrix $X = \{x_{ij}\}$, $\sum_{ij} x_{ij} = N$ describes the system *macrostate*. The number of microstates corresponding to X and differing in the number of particles is

$$W(X) = N!/\underset{ij}{M}\, x_{ij}\,! \tag{9}$$

The basic assumption of the approach is that every microstate is obtained at random and has no preferences over the other. To put it another way, $P(X)$ is the possibility of the macrostate X and it is proportional to $W(X)$. Then

$$\phi(X) = \ln P(X) \approx -\sum_{ij} x_{ij} \ln x_{ij} + c \; . \tag{10}$$

The expression (10) formally coincides with that for macrostate entropy. Together with analogies in the system structure this fact made it possible to introduce another assumption into [6]: a stable state $\dot{X}$ is obtained in the system corresponding to the maximal entropy (10). The generalization of [3] involving the introduction of probabilities ν_{ij}, different for every communication, results in the following form of generalized entropy

$$\phi(X) = \sum_{ij} x_{ij} \ln (v_{ij}/x_{ij}). \tag{11}$$

Now the task of modeling, i.e. determination of x_{ij}, reduces to maximization of the "natural" criterion $\phi(X)$ under constraints imposed on X. The authors of [3] classified constraints by three basic components of the system—origins, destinations, and communications with due account of the data on distribution for these components.

Thus, transportation constraints

$$\sum_j x_{ij} = P_i \,; \; \sum_i x_{ij} = Q_j \tag{12}$$

stand for data on distribution for origins P_i and destinations Q_j while the constraints on the average cost of subsystem allocation

$$\sum_j d_j \sum_i x_{ij} = ND \tag{13}$$

or on the mean movement time

$$\sum_{ij} t_{ij} \, x_{ij} = NT \tag{14}$$

stand for data on the mean values of distribution for destinations and communications, etc.

The values of v_{ij} can be determined by the values of communication characteristics. For an urban system, this distribution is referred to as the "settlement curve" $f_p(t)$ (Fig. 7) which represents the experimental data on the percentage of people moving with time costs t.

Hence, the problem of x_{ij} modeling is reduced to a formal mathematical programming problem

$$\max_{X \in \kappa} \phi(X) = \sum_{ij} x_{ij} \ln (v_{ij}/x_{ij}) \tag{15}$$

where the set of admissible values of κ is limited by the above constraints. For all combinations of the introduced constraints the problems are explicitly solved or solution algorithms have been proposed. Thus, solution of a problem with constraints (12) has the form

$$x_{ij} = \gamma_i \delta_j v_{ij} \tag{16}$$

where the parameters γ_i and δ_j are found by iterative substitution of (16) into (12) with subsequent "balancing".

This is the basic approach. Its extensions are generated by introduction of other constraints. Some of them are "quantitative" such as the recognition of inhomogeneity of population, residences, sites of work, diversity of communications, etc. The resulting problems are characterized by increasing scale, hence the respectively complex computations. Qualitatively different problems are engendered by the combinatorial-continuous constraints. For example, if in modeling the services allocation the required volume is below a certain threshold, then no allocation occurs.

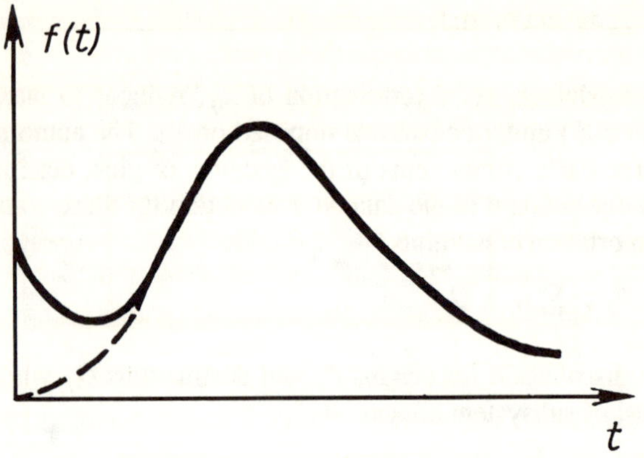

Figure 7.

At least two types of multipurpose movement models are possible: the successive choice of movement purposes or a random choice of the entire set of purposes.

A simple version of the first type of models for residential and service subsystems consists of two submodels [5]:

Residence | Service

$$\max \sum_{ij} x_{ij} \ln (v_{ij}/x_{ij}); \qquad \max \sum_{ij} y_{ij} \ln (m_{ij}/y_{ij}); \tag{17}$$

$$\sum_{j} x_{ij} = B_i + R_i; \qquad \sum_{i} y_{ij} = pR_j + rN_j + tB_j \tag{18}$$

$$N_j = \sum_{i} x_{ij} \leq (A_j - \alpha R_j) c^N; \qquad R_i = \sum_{j} y_{ij} \leq (A_i + \beta N_j) c^R \tag{19}$$

Every submodel maximizes the entropy of movement under constraints in the form of equalities (18) on the balance of levers and inequalities (19) on the area scarcity. Here B_i and R_i are the numbers of people employed in industries and services, N_i is the working population, and A_i is the area.

Given the fixed allocation of one subsystem, the "stable" allocation of the other depends on solution of the respective problem. A balance between the subsystems may be obtained by an alternating substitution of the stable allocations in the models of the other subsystem. In other words, the suggested model describes the interaction of subsystems through the sets of constraints (Fig. 8).

The idea of spatial interaction between subsystems is far from new, but it is the uniform "entropy" description that led to understanding what kind of equilibrium is determined in the model and made it possible to present the conditions for the algorithm's convergence to a unique equilibrium. Note that the convergence can be found experimentally but the uniqueness of the equilibrium cannot.

Recall now that all the proposed "entropy" models are simulation, behavioral ones, i.e. they are intended for determining the states (flows x_{ij}) with fixed exogenous signals (network structure, origins volumes P_i, destinations volumes Q_j, etc.) with an allowance for the population's behavior. The choice of P_i and Q_j values reducing, for

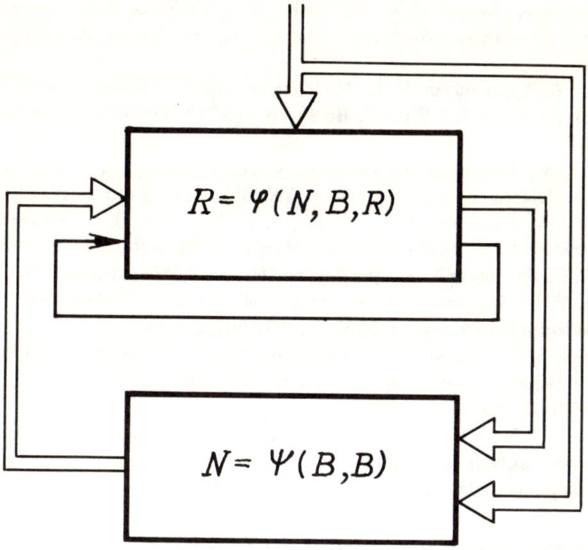

Figure 8.

example, the traffic remains outside the scope of the model and is generally exercised by a search of the alternatives. Advances in this direction call for the development of a normative-behavioral approach to model design. By definition two levels of criteria are introduced in such models: the upper one—normative, $f(x_{ij}, P_i, Q_j)$ and the lower one—behavioral. That is, the entropy $\phi(X)$ is considered under constraints such as (12). In other words the model is reduced, for example, to a problem

$$\min_{P_i, Q_j} f(x_{ij}, P_i, Q_j);\ (P, Q) \in \Omega \tag{20}$$

$$\max \phi(X) = \sum_{ij} x_{ij} \ln (v_{ij}/x_{ij})$$

$$\sum_j x_{ij} = P_i;\ \sum_i x_{ij} = Q_j\ .$$

The solution for the lower level is unique with fixed P_i and Q_j, i.e. $x_{ij} = x_{ij}(P_i, Q_j)$ (16), hence, (18) is reduced to

$$\min_{P_i, Q_j} f\,[x_{ij}(P_i, Q_j), P_i, Q_j] = \psi(P_i, Q_j) \tag{21}$$

The developed methods and algorithms have been successfully employed in many urban design problems.

REFERENCES

1. Ayres, R. U., *Technological Forecasting and Long-Range Planning*, McGraw-Hill, New York, 1966.

2. Berezovsky, B. A., Kossov, O. A., Mashinsky, A. L., Yakimetz, V. N., *Generation of Sets of Higher Plants for a Closed-Cycle Biotechnological Life Support System*, Microbioprom Publishers, Moscow, 1976.
3. Imelbayev, Sh.S., Shmulyan, B. L., Modeling of Stochastic Communication Systems (Addenda to the Russian Edition of A. J. Wilson's Book), pp. 170-233, Nauka Publishers, Moscow, 1978 (in Russian).
4. Kossov, O. A., Travkin, S. I., Yakimetz, V. N., Gloustchenko, A. S., The Condition of Multiparametric Balance in System Generation, VNIISI Preprint, Moscow, 1978 (in Russian).
5. Popkov, Yu. S., Posokhin, M. V., Gutnov, A. E., Shmulyan, B. L., *Systems Analysis and Urban Development Problems*, Nauka Publishers, Moscow, 1983 (in Russian).
6. Wilson, A. G., *Entropy in Urban and Regional Modeling*, Pion, London, 1970.
7. Yakimetz, V. N., Generation of Lexicographically Ranked Morphological Versions of a System, *VNIISI Collected Papers*, Issue 1, Moscow, 1981 (in Russian).
8. Yemelyanov, S. V., Korovin, S. K., Application of an Error-Closing Control Principle for Expansion of a Set of Feedback Types, *Papers of the USSR Academy of Sciences*, vol. 258, no. 5, pp. 1070-1074, 1981 (in Russian).
9. Yemelyanov, S. V., Korovin, S. K., The Design Principles and Basic Properties of Closed-Loop Dynamic Systems with Different Types of Feedback, in *Dynamics of Inhomogeneous Systems*, VNIISI Workshop Proceedings, pp. 5-27, 1982 (in Russian).

DISCUSSION OF KOSSOV'S PRESENTATION

Umpleby

What is the purpose—the motivation—which lies behind and drives your work?

Kossov

The basic motivation is the search for new methods of resolving practical tasks including non-traditional ones. Thus, in contemporary metallurgical, oil, biotechnological and other branches of industry there is a well-known movement for the use of more intensive technological processes in which the control parameters of objects change rapidly and uncontrollably, being located at the edge of system stability. On the basis of the ideas set forth in the first part of the report, certain problems of that type have been resolved—algorithms have been proposed for structuring new control systems which work effectively in industry.

I will give one example to illustrate the possibility of the methods mentioned in the second part. We were asked to form 3-5 variant sets, each containing 5-8 species, from a list of 35 species for inclusion in a closed biotechnical life-support system for future space flights. Analysis showed that there were 25 × 106 different variants. To compare and evaluate all of them and make a selection would be enormously labor-intensive. Our method enabled us to form only those variants which satisfied the demands of the multi-criteria evaluation, without having to sort through all of them.

Finally, the entropy approach has been reduced in our Institute to a set of computer programs. These are widely used by architectural and planning organizations in Moscow and other cities for evaluating and comparing alternative long-term development programs and concrete planning decisions. We have even used these methods to alter the time of starting the working day of many enterprises to reduce the peak load on urban transport systems.

CHAPTER

7

RECONSTRUCTABILITY ANALYSIS: AIMS, RESULTS, OPEN PROBLEMS†

George J. Klir and Eileen C. Way

The term "reconstructability analysis" has evolved as a generic term for all aspects of a methodology whose aim is to deal with problems associated with the relationship between systems perceived as wholes and their various subsystems (i.e. parts of the wholes). This article is a state-of-the-art report on reconstructability analysis (RA). It contains a description of problems that are studied under RA, a summary of main results obtained within RA, an outline of some important issues associated with RA that require further research, and an up-to-date bibliography of RA.

After a general discussion of the issues associated with the whole/part relationship and an overview of the aims of RA, the article describes results obtained from two problems that are subjects of study in RA. One of the problems, called an identification problem, is concerned with inferences about an unknown overall system

†This work was supported by National Science Foundation under Grants ECS 8217103 and INT-8219423. Reprinted with permission from *Systems Research*, Vol. 2, No. 2, pp. 141-163, 1985, Pergamon Journals Ltd.

from a given set of its subsystems. A deductive inference from the given subsystems results in the set of all overall systems which yield those very subsystems; it is called a reconstruction family. The ability to identify the true overall system is measured by an identifiability quotient that is inversely proportional to the relative size of the reconstruction family. A selection of one overall system from the reconstruction family requires an inductive inference. It is argued that the selected system should be unbiased and a simple procedure for determining it (called a join procedure) is described.

The second problem, called a reconstruction problem, is concerned with the determination of such subsystems, as small as possible, from which the overall system can be reconstructed to an acceptable degree of approximation. The approximation is measured by an information distance between the given and reconstructed overall systems. The reconstruction problem involves a number of computational difficulties. Some approaches to overcoming these difficulties are discussed. The article also contains a description of simulation experiments which contributed toward the determination of some fundamental characteristics of RA for the reconstruction problem.

1 RECONSTRUCTABILITY ANALYSIS: AN OVERVIEW

Reconstructability analysis (RA) is an organized collection of ideas, concepts, principles and methods whose primary purpose is to provide systems investigators with a convenient framework and methodological tools for discussing and dealing with the various questions associated with a key problem area of systems inquiries—the relationship between wholes and parts or, more specifically, between overall systems and their various subsystems. The aim of this article is to describe problems that are subjects of study in RA, present a "state-of-the-art" report on RA, outline some important open problems associated with RA and provide the reader with an up-to-date bibliography of RA.

RA emerged from ideas formulated by Klir in 1976 [67]. These ideas were sparked primarily by some work of Ashby. In the mid-1960s, Ashby suggested the concept of cylindrance for many-dimensional relations on finite sets and developed an algorithm to determine whether a given relation of dimension n can be reconstructed from all its projections of a particular dimension $k < n$. The reconstruction, if possible, is accomplished by the set intersection of extensions (cylinders) of all the k-dimensional projections of the given relation [1,2]. Later, Ashby and Madden discussed the inverse problem, i.e. the problem of identifying an unknown relation among n nominal scale variables from relations among subsets of k of these variables ($k < n$) [86]. Although they were well aware of the significance of this problem, no methods for dealing with it resulted from their work. Another aspect of Ashby's work—his use of information theory as a general tool for analyzing systems [3-6]—was also influential in forming the initial ideas of RA.

Since the initial inception of RA in 1976, many conceptual, mathematical and computational results have been obtained, which have made it possible to develop computer software for dealing with the various issues of the system/subsystem relationship. RA itself has also been tested by extensive simulation experiments and its applicability has been explored in some areas, including computer performance evaluation, ecological modeling, agriculture, medicine and social science research [19,22,32,55,56,71,75,78,90,94,96,98,99,101].

RA is predominantly based upon principles of information theory. They are formulated, for the most part, within classical set theory and probability theory, but attempts have recently been made to extend the formulation to the more general framework of fuzzy set theory [45]. Thus far, RA has been extended only to one area of fuzzy set theory, which is referred to as possibility theory [45,102].

The primary purpose of fuzzy set theory is to more closely capture the nature of human reasoning. It is based on the notion that people do not reason in two-valued logic but rather they use vagueness and imprecision in thinking and decision making. This allows them to make task-oriented trade-offs between complexity and precision. When the scope of an unmanageable task cannot be reduced without losing relevance, then a reduction of precision is one way, perhaps the most natural one, to make the task manageable. Classes of objects whose membership is gradual rather than abrupt are used to represent this imprecision. These are called fuzzy subsets and are characterized by degrees of membership of elements of some universal set that range between 0 and 1 inclusive.

The most fundamental concept in RA is that of a *system*. The concept of a system means different things in different contexts: a system can mean a method or way of doing something, a theory, or that which a theory is about [88]. In RA, the term 'system' is viewed as some kind of characterization of the constraint associated with a set of variables. In this characterization, the concept of fuzzy measures is essential [45], particularly some special classes of fuzzy measures such as probability and possibility measures. It is usually convenient to express measures in these special classes by the respective distributions (e.g. probability or possibility distributions) defined on the set of states of the variables involved (it is assumed that the set of states is finite).

Formally, a *system* is defined in RA as the sextuple

$$F = (V;\mathcal{V};s;A;Q;f), \tag{1}$$

where $V = \{v_i \mid i \in N_n\}(N_n = \{1,2,\ldots,n\}\dagger)$ is a set of *variables*; $\mathcal{V}$ is a family of finite *state sets;s*: $V \rightarrow \mathcal{V}$ is an onto *assignment function* by which one state set from $\mathcal{V}$ is assigned to each variable in V;

$$A = s(v_1) \times s(v_2) \times \cdots \times s(v_n)$$

is the set of all *overall states* (n-tuples) defined for the variables; Q is a set of real numbers which includes 0, usually the set [0,1]; $f: A \rightarrow Q$ is a function by which the *constraint* of the variables is characterized (e.g. a probability or possibility distribution function). When some of the variables are declared as input variables, function f characterizes a conditional constraint (e.g. conditional probabilities) in which states of the input variables represent the conditions.

†N subscripted by a positive integer is always used in this article to denote the set of positive integers from 1 through the value of the subscript.

It is understood that the variables in (1) may be direct images of some real-world attributes (each observed with some resolution), indirect images of the attributes (such as lagged variables), or merely some abstract entities with no apparent real-world meaning (e.g. postulated internal variables or variables in purely mathematical systems). These distinctions between variables, which are necessary for characterizing dynamical aspects of systems, are essential in overall systems investigations. Although they are relevant to questions regarding the interpretations of the results of RA, they are of no significance to RA itself. For this reason, the distinctions between variables have not been considered during the developmental stage of RA; their study in the context of various applications is outside the scope of RA and, consequently, it is not discussed in this article.

RA deals with sets

$$S = \{{}^kF = ({}^kV; {}^k\mathcal{V}; {}^ks; {}^kA; {}^kQ; {}^kf) \mid k \in N_q\} \tag{2}$$

of systems ${}^kF$ defined by the form (1). Such sets are referred to as *structure systems*. In general, the sets ${}^kV\,(k \in N_q)$ in (2) may overlap in any arbitrary way. Systems ${}^kF$ that are contained in S are usually called *elements* of the structure system. It is assumed that all elements of a structure system are compatible in the sense that constraints among their variables are characterized by the same type of measures.

If elements ${}^kF$ of a structure system defined by (2) are *subsystems* of a system given by (1), then ${}^kV \subset V$ for all $k \in N_q$; consequently, states in sets ${}^kA$ are appropriate substates of those in A, and functions ${}^kf$, defined on these substates, are appropriate projections of f (determined by the type of measure employed in characterizing the constraint of the variables in V). The concepts of a substate and a projection are essential for RA and require formal definitions.

For each overall state (n-tuple) $\alpha = (\alpha_i \mid i \in N_n) \in A$ of a system defined by (1) and for each state $\beta = (\beta_j \mid j \in J \subset N_n)$ associated with variables in set ${}^kV = \{v_j \mid j \in J \subset N_n\}$, β is called a substate of α if and only if $\beta_j = \alpha_j$ for all $j \in J$. Let $\beta < \alpha$ denote that β is a substate of α.

In most papers on RA, $[f \downarrow {}^kV]$ denotes the projection of f which involves only variables in set ${}^kV \subset V$. It is a function

$$[f \downarrow {}^kV]: \underset{v_j \in {}^kV}{\times} s(v_j) \rightarrow Q$$

such that

$$[f \downarrow {}^kV](\beta) = a(\{f(\alpha) \mid \alpha > \beta\}),$$

where a is an aggregation function required by the measure used. For probabilistic systems,

$$[f \downarrow {}^kV](\beta) = \sum_{\alpha > \beta} f(\alpha); \tag{3}$$

for possibilistic systems,

$$[f \downarrow {}^kV](\beta) = \max_{\alpha > \beta} f(\alpha). \tag{4}$$

Naturally, the concepts of wholes and parts are essential to RA and they have been termed overall systems and subsystems, respectively. The status of a system as either an overall system or a subsystem is not absolute. Any given system can assume the status of a subsystem in one context and the status of a supersystem in another context. We may say, more poetically, that 'a part is a whole in a role' (R. Glanville), but also that a whole, in fact, is a part in a role. This duality makes it possible to represent each overall system by a hierarchy of collections of subsystems, ie. by a structure system whose elements are also represented by structure systems, whose elements are also..., etc., up to elements consisting of single variables. Where we set the boundaries for what is to be considered the overall system and what are its subsystems depends on the purpose and constraints of our inquiry.

We are now in a position to begin to discuss the results of the research done on RA. Part of these results has, of course, been the conception and formalization of several of the above definitions, as well as the very concept of RA itself. Deriving the correct formal representations of these concepts is vital in order to insure a complete and consistent framework within which the various issues surrounding the relationship between wholes and parts can be clearly formulated as systems problems and studied accordingly.

Two important steps in the initial stage of developing RA were: (i) the recognition that there are two complementary problems involved, and (ii) an adequate formulation of these problems [23,27]. In one of the problems, a structure system of type (2) is given whose elements ${}^kF$ are viewed as subsystems of an unknown overall system F of type (1), i.e. V in (1) is the set union of sets ${}^kV\,(k \in N_q)$ in (2). The aim is to derive from the subsystem ${}^kF$ as much knowledge as possible regarding the unknown overall system F. This problem has been called the *identification problem.*

The second problem is based on the assumption that an overall system is given. The aim in this problem is to determine which sets of subsystems are adequate for reconstructing the given overall system, to an acceptable degree of approximation, solely from the knowledge embedded in the subsystems. This problem has been referred to as the *reconstruction problem.*

2 IDENTIFICATION PROBLEM

The identification problem, when defined within the context of RA, is the problem of identifying specified properties of an unknown overall system from the knowledge of corresponding properties of some given set of its subsystems. The identification problem is thus concerned with inferences from a given structure system of the form (2) about an unknown overall system of the form (1).

Why should an investigator use only subsystems rather than the whole system of his interest? The main reason is that he is frequently dependent upon relevant results, such as experimental data or systems, developed by other investigators. Interdisciplinary inquiries, for example, often involve isolated results obtained by specialists in various disciplines. Moreover, even if the inquiry is self-contained, it is often technically impossible to simultaneously measure all the variables of the overall system. The best the investigator can do in such circumstances is to study the overall system in terms of some of its subsystems. The choice of the subsystems is based in such cases solely on practical considerations and does not imply that the overall system is adequately represented by the subsystems. One aim of the identification problem is to determine the extent to which the subsystems do adequately represent the overall system.

2.1 Reconstruction families

The identification problem in the context of RA has been motivated by the realization that for any given structure system S of the form (2) there exists a class of hypothetical overall systems of the form (1). Each of them, for example system F, is compatible with the given structure system in the sense that for each ${}^kF \in S$, F and ${}^kF$ have a system-subsystem relationship. Formally, this means that for each $k \in N_q$, F satisfies the following conditions: (i) ${}^kV \subset V$; (ii) ${}^k\mathcal{V} \subseteq \mathcal{V}$ such that ${}^ks$ is onto; (iii) ${}^ks : {}^kV \to {}^k\mathcal{V}$ such that ${}^ks(v_i) = s(v_i)$; (iv) ${}^kA = \times s(v_i)$, where $v_i \in {}^kV$; (v) ${}^kQ = Q$; (vi) ${}^kf = [f \downarrow {}^kV]$. This class of hypothetical overall systems of S is called a *reconstruction family* of S.

Determining the reconstruction family of a given structure system is a part of the identification problem. The six conditions required for a system-subsystem relationship are crucial for a mathematical formulation of this aspect of the problem. The formulation was developed for two types of systems—probabilistic and possibilistic systems.

For probabilistic systems the reconstruction family is completely described by the solution of the set of simultaneous linear algebraic equations

$$ {}^kf({}^k\beta) = \sum_{\alpha > {}^k\beta} f(\alpha) \tag{5}$$

for all $k \in N_q$ and all ${}^k\beta \in {}^kA$; these equations express the fact that the given marginal probability distributions $({}^kf({}^k\beta) \mid {}^k\beta \in {}^kA)$ in the structure system $(k \in N_q)$ are projections [in the sense of equation (3)] of an unknown overall probability distribution $(f(\alpha) \mid \alpha \in A)$. They are usually underdetermined and are always constrained by the inequalities

$$ f(\alpha) \geq 0 \tag{6}$$

for all unknowns $f(\alpha)$, $\alpha \in A$.

For more complex structure systems, reconstruction families may be characterized by several free parameters and these may be mutually interdependent. The solution process becomes then quite involved and the use of a formal method and the computer is unavoidable. Since, in spite of the many methods described in the literature for dealing with linear algebraic equations, no method was available for solving this specific problem, we developed our own matrix method; it is described in [27]. An alternative method was later proposed by Jones [63,64]. These methods can be directly applied to a more general identification problem formulated by Bard [7,8]. In this problem, probabilities of arbitrary events defined on a state set are given, rather than probabilities of events imposed by the projections involved. Events are viewed in this context as subsets of the state set A.

For possibilistic systems, the simultaneous equations have the form

$$\max_{\alpha > {}^k\beta} f(\alpha) = {}^k f({}^k\beta) \tag{7}$$

for all $k \in N_q$ and ${}^k\beta \in {}^kA$, and are again constrained by (6). This problem was adequately solved by converting it to a special form of fuzzy relation equations [59,60]. It was proved that each reconstruction family can be fully described in terms of the maximum solution (unique) and all minimal solutions of these equations. On the basis of this theoretical result, an efficient procedure was developed for determining reconstruction families for possibilistic structure systems. The details of this approach are beyond the scope of this paper; they can be found in [60].

2.2 Consistency of Structure Systems

It is now established that, regardless of the type of measures employed for characterizing systems constraints, a necessary (but not sufficient) condition for the reconstruction family to be nonempty is that the elements of the given structure system of the form (2) be *locally consistent* in the sense that the equation

$$[{}^j f \downarrow {}^j V \cap {}^k V] = [{}^k f \downarrow {}^j V \cap {}^k V] \tag{8}$$

is satisfied for all pairs $j, k \in N_q$. This condition is not sufficient because there exist structure systems that are locally consistent and, yet their reconstruction families are empty. Such structure systems are called *globally inconsistent.*

Local inconsistencies in a structure system may be (and usually are) caused by the fact that the behavior functions associated with its elements are only estimates, each derived from limited experimental data. Hence, local inconsistencies are a result of our ignorance and, consequently, it is meaningful to try to resolve them by a suitable method; this issue is discussed later in this paper. A global inconsistency, on the other hand, is more serious. Its meaning is that the structure system is ill-conceived; it is a mathematical artifact that has no meaning in the real world. Excellent examples of globally inconsistent wholes whose elements are locally consistent can be found in the world of graphic arts. We have in mind, for example, some drawings by M. C. Escher (such as his lithograph *Belvedere*) and, particularly, the many drawings by the Swedish artist Oscar Reutersvärd, which are referred to as impossible figures or perspective japonaise.†

2.3 Identifiability Quotient

In addition to the determination of reconstruction families, the concept of an *identifiability quotient* has been introduced, by which the ability to identify a unique overall system from the given structure system (assumed to be consistent) is measured [27,60]. Its value should be 1 if and only if the reconstruction family consists of only

†Oscar Reutersvärd, *Omöjlige figurer.* Doxa, Arlöv (Sweden) (1982).

one overall system or 0 if we have no information about the true overall system (the latter case obviously cannot happen in the identification problem because the elements of the given structure system always contain some information). In the general case, the identifiability quotient should measure the relative size of the reconstruction family; the smaller the size, the larger the value of the identifiability quotient.

For possibilistic systems, a well justified identifiability quotient and a method by which it can be calculated very easily has been developed and tested [60]. This identifiability quotient is a powerful tool for quick evaluations of structure systems, particularly those for which the determination of the reconstruction family is too costly or even intractable. For probabilistic systems, an adequately justified identifiability quotient remains an open problem.

2.4 Unbiased Reconstruction

Given a structure system, all members of its reconstruction family are hypothetical overall systems of the structure system, but we do not know which of them is the true overall system. If we must select one of these hypothetical overall systems, we need some justification for choosing one of them over the others. That is, we must resort to some justifiable principle of inductive inference. First of all, it is essential to require that the chosen overall system be unbiased, i.e. implied solely by the information (constraints) contained in the structure system. Let such a unique member of the reconstruction family be called an *unbiased reconstruction.* To determine the unbiased reconstruction, it is required that all but no more information than available in the structure system about the overall system be used. Hence, the unbiased reconstruction is required to be maximally non-committal with regard to all matters except the partial constraints involved in the structure system.

When the probability measure is used to characterize the constraints, the unbiased reconstruction is represented by that overall system in the reconstruction family whose probability distribution has larger entropy (in the Shannon sense) than any other system in the family. The unbiased reconstruction is thus the reconstruction with maximum entropy subject to the constraints expressed by the given structure system (and, possibly, also subject to additional constraints representing valid background information). This fact is directly derivable from a more general *principle of maximum entropy* [31,62]. Furthermore, the maximum entropy reconstruction is unique.

When a choice of a single system from the reconstruction family is necessary, we are justified in choosing the system on the basis of the maximum entropy principle by at least four diverse reasons:

1. Jaynes, Christensen and others have shown that the maximum entropy probability distribution of the overall system is the only unbiased distribution, i.e. the only distribution which takes into account all available information (constraints in the given reconstruction hypothesis), but no additional (unsupported) assumptions (biases) [31,62]. Thus, to accept any other probability distribution would mean that we are including additional and possibly erroneous information; in other words, we are biasing our estimate of the probabilities.

2. As proved by Shore and Johnson, maximizing any functional but entropy leads to inconsistencies unless the functional has the same maxima as entropy [93].

3. Furthermore, Jaynes has shown that based on strictly combinatorial calculations, the maximum entropy probability distribution is the most likely distribution of all the distributions associated with the reconstruction family [61].

4. Finally, we have developed an argument that it is not possible to design and build a real-world system by a set of subsystems characterized in a probabilistic way such that its overall probability distribution is different from the maximum entropy reconstruction [27].

It is also well known that the principle of maximum entropy (and its generalized form—the principle of maximum cross-entropy) operationalizes *inductive inference* for systems whose constraints are expressed by the probability measure [31,93].

2.5 Join Procedure

Determining the maximum entropy reconstruction requires an appropriate method for dealing with the nonlinear constraint optimization problem. For the basic identification problem (i.e. one which does not involve any background information), it is now well established that the maximum entropy reconstruction can be determined by a computationally simple and straightforward procedure, without actually solving the optimization problem [27]. We developed this procedure, referred to as the *join procedure*, in the early stages of our research on RA. It is based on the relational operation, by which a ternary relation $R \subset I \times J \times K$ is formed from two binary relations $R_1 \subset I \times J$ and $R_2 \subset J \times K$ according to the following definition:

$$R = \{(\alpha,\beta,\delta) \mid (\alpha,\beta) \in R_1 \text{ and } (\beta,\delta) \in R_2\} . \tag{9}$$

If elements of the relations are characterized by their probabilities $f(\alpha,\beta,\delta)$, $f_1(\alpha,\beta)$ and $f_2(\beta,\delta)$, then

$$f(\alpha,\beta,\delta) = f_1(\alpha,\beta) \cdot f_2(\delta \mid \beta) , \tag{10}$$

where $f_2(\delta \mid \beta)$ is the conditional probability of δ given β. This operation is called a *relational join* and it is usually denoted by an asterisk; thus, $f = f_1 * f_2$.

Assume that the relational join is applied to two elements of a structure system whose sets of variables and probability distribution functions are, respectively, ${}^jV$, ${}^kV$ and ${}^jf$, ${}^kf$. Then, the domains ${}^jA$ and ${}^kA$ of the respective functions must be converted into the forms $I \times J$ and $J \times K$, respectively, where

- I is the set of all overall states of variables which participate only in the first element, i.e. variables in the set ${}^jV - {}^jV \cap {}^kV$;
- J is the set of all overall states which participate in both of the elements, i.e. variables in ${}^jV \cap {}^kV$;
- K is the set of all overall states of variables which participate only in the second element, i.e. variables in the set ${}^kV - {}^jV \cap {}^kV$.

When there is no shared set of variables between the subsystems, then the join operation degenerates into Cartesian product of the relevant state sets and the product of relevant basic probabilities.

Given a structure system, the relational join is applied repeatedly to pairs of elements. In each of its applications, two elements merge (join) into a larger element. Assume that the join is always performed so that the result of previous joins enters as the second function. The evaluation is completed when all elements merge into one overall system provided that the structure system involved does not contain loops [24,56,75,84]. (These so-called *loopless structure systems* are defined in the Appendix.) For structure systems with loops, the join procedure has to be repeated iteratively to converge to the maximum entropy reconstruction; more details on this issue can be found in [27].

In order to develop a similar procedure for possibilistic systems, we had to derive a *possibilistic measure of uncertainty/information* first. We proposed such a measure intuitively in 1981 and proved later [57] that it possesses possibilistic counterparts of all properties of the Shannon entropy and, at the same time, is a natural generalization of the Hartley information [51]. To describe it, let ${}^nP$ denote the set of possibility distributions associated with n outcomes and let $f = (\phi_1, \phi_2, \ldots, \phi_n)$ denote a particular possibility distribution in ${}^nP$. Then, the possibilistic uncertainty associated with each $f \in {}^nP$ is expressed by a function

$${}^nU : {}^nP \rightarrow R^+$$

such that

$${}^nU(f) = \frac{1}{l_f} \sum_{k=1}^{r-1} (l_{k+1} - l_k) \log_2 |{}^nc(f, l_{k+1})|, \qquad (11)$$

where $l_1, l_2, \ldots, l_r$ denote distinct levels associated with f; it is assumed that

$$l_1 = 0, 0 < l_2 < l_3 \ldots < l_r (= l_f).$$

Symbol ${}^nc$ in (11) denotes a function

$${}^nc : {}^nP \times R^+ \rightarrow P(N_n)$$

such that

$${}^nc(f, l) = \{ i \mid \phi_i \geq l \}, \; l_f = \max_i \phi_i,$$

and $|{}^nc(f, l)|$ denotes the cardinality of ${}^nc(f, l)$; ${}^nc(f, l)$ is usually called the l-cut of f [45].

The U uncertainty can be viewed as the mean of uncertainties (each expressed by the Hartley measure) for all l cuts associated with the r-tuple of levels $(l_1, l_2, \ldots, l_r)$ of f, weighted by the differences in the levels. Observe that (11) can also be written as

$${}^nU(f) = \frac{1}{l_f} \int_0^{l_f} \log_2 |{}^nc(f, l)| \, dl . \qquad (12)$$

Possibilistic uncertainty expressed by (11) or (12) can be directly used for developing a principle of inductive inference, analogous to the principle of maximum entropy, for systems characterized by the possibility measure. The join procedure

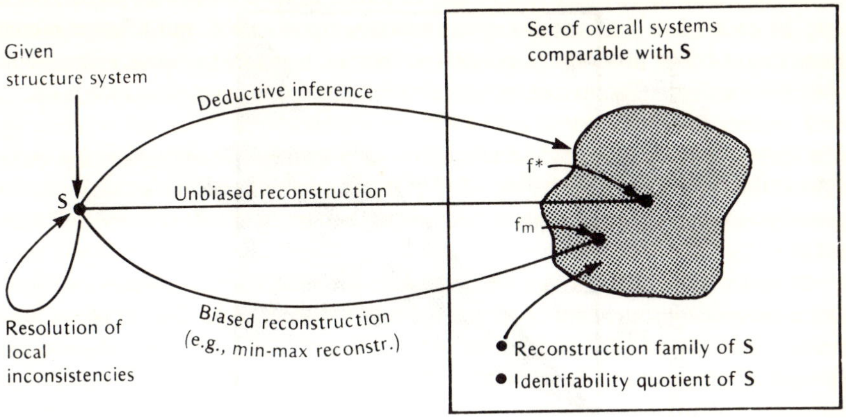

Figure 1 Summary of basic issues involved in the identification problem.

similar to the one described above (for probabilistic systems) was developed for possibilistic systems to determine the maximum uncertainty reconstruction [30]; it just replaces (10) with

$$f(\alpha,\beta,\delta) = \min[f_1(\alpha,\beta), f_2(\beta,\delta)].$$

Observe that no conditional possibilities are needed here. It is also known that no iterations are needed for possibilistic structure systems with loops to obtain the maximum uncertainty reconstruction [30]. It is thus computationally more efficient to handle possibilistic systems in RA than their probabilistic counterparts.

A generalized join procedure for probabilistic systems was developed by Jones [64,65]. It determines the unbiased reconstruction even when probabilities are not known for all states of the elements of the given structure system. When they are known the procedure identifies and utilizes only an irredundant subset of them. This procedure is also applicable to the generalized reconstruction problem formulated by Bard [7,8].

The basic issues involved in the identification problem are summarized in Fig. 1. Symbol f_m denotes some kind of a biased reconstruction, for example the so-called min-max reconstruction proposed in Section 5.

3 RECONSTRUCTION PROBLEM

The aim of the reconstruction problem is to determine from among all relevant structure systems those which allow us to reconstruct a given overall system to an acceptable degree of approximation. Thus, in this problem, we already have the overall system and we want to find a way to break it into smaller parts while retaining as much information as possible. In general, the parts may overlap (i.e. they may share some variables).

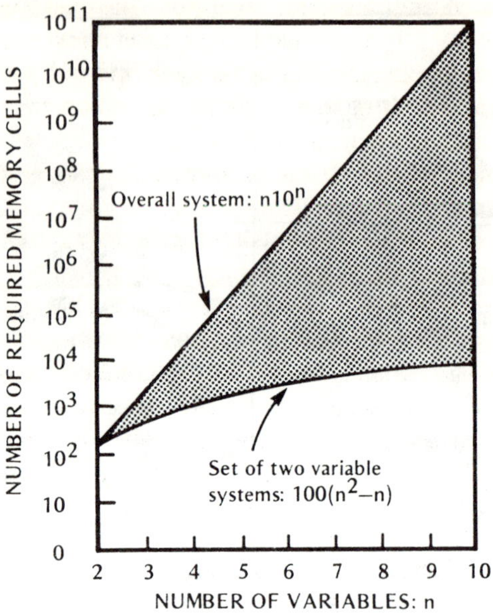

Figure 2 Comparison of memory requirements for an overall system and the associated structure system consisting of all two-variable subsystems ($k = 10$).

Why is it desirable to represent overall systems by collections of their subsystems? There are several reasons. One of them is concerned with observation or measurement: it is often impossible or, at least, impractical to observe (measure) simultaneously all variables that are considered relevant for some purpose of inquiry. Another reason is connected with complexity and that, in turn, is connected with manageability of the system under consideration. For example, it is easier to monitor four sets of three physiological variables of a patient independently of each other than to monitor all of the variables involved (say 10 of them) simultaneously during a difficult surgery, when appropriate decisions and actions must be made quickly.

One aspect of systems manageability is expressed in terms of the size of computer memory required to store the system. Consider, for example, n variables, each of which has k states. When dealing with the overall system of these variables, nk^n memory cells, each of which can store any one of k states, must be made available for storing states of the system. On the other hand, when a structure system consisting of all subsystems with two variables is used, the number of memory cells that are needed for the same purpose is $k^2n(n-1)$. This number grows with increasing values of k and n at a considerably lower rate than that of nk^n, as illustrated in Fig. 2 for $k = 10$. If the structure system contained only some of the two-variable subsystems, the comparison would be even more favorable for the subsystem representation of the overall system. Furthermore, if we only stored states of the subsystems that are needed to form a linearly independent set of equations (5), as inherent in the method proposed by

Jones [63], the structure system would be even more advantageous when compared with the overall system.

Practical reasons related to problems of measurement, manageability, efficiency etc. are not the only reasons why it is desirable to represent overall systems by appropriate structure systems. In systems inquiries, structure systems have a more fundamental role. When properly justified, a structure system provides the investigator with some knowledge that is not available, at least explicitly, in the corresponding overall system. For example, the configuration of the subsystem in the structure system gives us information about causal relationships, the significance of individual variables, the strength of the dependencies among them, etc. This additional knowledge may help the investigator to answer certain questions associated with the inquiry or, more generally, may help him to develop a better insight into the associated problem.

3.1 Reconstruction Hypotheses

The question of which structure systems are relevant to the reconstruction problem led us to the formulation of the concept of a *reconstruction hypothesis*: it is a structure system whose elements are subsystems of the given overall system and which, in addition, satisfies the conditions

$$\bigcup_{k \in N_q} {}^{k}V = V \tag{13}$$

and

$$(\forall j, k \in N_q)({}^{j}V \subseteq {}^{k}V \Rightarrow j=k), \tag{14}$$

referred to, respectively, as the *covering condition* and the *irredundancy condition* [24]. The covering condition guarantees that all variables of the given overall system are included in the hypothetical structure system. The aim of the irredundancy condition is to exclude such structure systems that contain at least one element which is a subsystem of another element. Such an element, which is a subsystem of the overall system within the considered structure system, does not contribute any information that is not included in the larger element.

If we also wanted to consider structure systems with no information about some variables of the given overall system, we would disregard the covering condition. Let structure systems whose elements are subsystems of the overall system and which satisfy only the irredundancy condition (14) be called *extended reconstruction hypotheses* [74,83]. Although the reconstruction problem has been predominantly studied within the context of reconstruction hypotheses, virtually all results can be easily extended to generalized reconstruction hypotheses.

Given an overall system F of the form (1), each reconstruction hypothesis is fully characterized by two entities: (i) a family of subsets of the variables involved in the overall systems [subsets ${}^{k}V$ in (2)], and (ii) properties of these subsets of variables (components ${}^{k}\mathcal{V}$, ${}^{k}s$, ${}^{k}A$, ${}^{k}Q$, ${}^{k}f$ for each subset ${}^{k}V$). The selection of a family of subsets of the variables involved is independent of the given overall system. However, when a family is selected, the properties of each subset of variables are uniquely derivable from the overall system. When these properties are disregarded, the family of subsets of variables characterizes only the structure of a prospective reconstruction hypothesis. It become an actual reconstruction hypothesis when the variables involved in it are interpreted as variables of a particular overall system. It is often more meaningful to deal with structures of reconstruction hypotheses rather than the actual reconstruction hypotheses of some specific overall system. When they are not restricted in any way except for the covering and irredundancy conditions, the structures are usually referred to in the RA literature as *general structures* or *G-structures.* Let structures that represent extended reconstruction hypotheses (those not required to cover all variables) be called *extended general structures* or G^+-structures. Observe that G-structures are actually hypergraphs of a special kind, namely, hypergraphs that satisfy the irredundancy condition (14)[9]; they are obviously, also, a special kind of G^+-structures.

3.2 Information Distance

Given a set of reconstruction hypotheses, we need to evaluate how much information about the original overall system is contained in each of them. Since such evaluations are based on the overall systems reconstructed from the various hypotheses, we must insure that the reconstruction method used utilizes all information available in each of these hypotheses. But, at the same time, we want to be sure that no additional and unsupported (i.e. biasing) information is used in deriving the reconstructed system. One method that allows us to determine the unbiased reconstruction of a structure system (assuming no background information is involved) is the join procedure. The arguments supporting this choice are the same as those used in the identification problem (Section 2.4). After the unbiased reconstruction (i.e. the reconstructed overall system) is determined, it must then be compared with the actual (given) overall system. We realized that it was essential to use an *information distance* in this comparison, one which would measure the difference between the amount of information contained in the given overall system and the fraction of the information preserved in the reconstruction hypothesis. For probabilistic systems this distance is naturally expressed by the well-known formula [51]

$$d(\mathbf{f},\mathbf{f}_s) = \frac{1}{\log_2 |A|} \sum_{\alpha} f(\alpha) \log_2 \frac{f(\alpha)}{f_s(\alpha)}, \tag{15}$$

where the constant $1/log_2|A|$ ($|A|$ denotes the cardinality of A) is a normalizing factor to yield the property

$$0 \le d(\mathbf{f},\mathbf{f}_s) \le 1.\dagger \tag{16}$$

For possibilistic systems, an analogous distance was formulated in terms of the possibilistic measure of uncertainty and information [58]; for normalized $\mathbf{f}$, $\mathbf{f}_s$:

---

†$\mathbf{f}$ and $\mathbf{f}_s$ denote the probability or possibility distributions of the given overall system and the overall system reconstructed from the reconstruction hypothesis S, respectively.

$$d(\mathbf{f},\mathbf{f}_s) = \frac{1}{\log_2|A|}\int_0^1 \log_2\frac{|{}^n c(\mathbf{f}_s,l)|}{|{}^n c(\mathbf{f},l)|}\,dl\,. \tag{17}$$

Although these information distances are applicable for comparing a given overall system with unbiased reconstruction from any of its reconstruction hypotheses, they are not applicable to all pairs of generative systems defined on the same set of states. A *generalized information distance*

$$D(\mathbf{f}_1,\mathbf{f}_2) = \mathrm{d}(\mathbf{f}_1,\mathbf{f}_1 \vee \mathbf{f}_2) + \mathrm{d}(\mathbf{f}_2,\mathbf{f}_1 \vee \mathbf{f}_2) \tag{18}$$

was proposed for possibilistic systems, where

$$[f_1 \vee f_2](\alpha) = \max_{\alpha\in A}\,[f_1(\alpha), f_2(\alpha)] \tag{19}$$

and d is defined by (17). It was proven that D is a metric information distance and is applicable to all pairs $\mathbf{f}_1,\mathbf{f}_2$ of possibility distributions defined on the state set A [58]. For probabilistic systems, a similar distance was proposed.

$$D(\mathbf{f}_1,\mathbf{f}_2) = \mathrm{d}\left(\mathbf{f}_1, \frac{\mathbf{f}_1+\mathbf{f}_2}{2}\right) + \mathrm{d}\left(\mathbf{f}_1, \frac{\mathbf{f}_1+\mathbf{f}_2}{2}\right), \tag{20}$$

where d is defined by (15) and

$$\frac{f_1+f_2}{2}(\alpha) = \frac{f_1(\alpha)+f_2(\alpha)}{2} \tag{21}$$

for all $\alpha \in A$ [58].

3.3 Refinement Ordering

The most difficult issue in the reconstruction problem is the search through the set of all relevant reconstruction hypotheses. To make the search through this set meaningful as well as orderly and efficient, it is essential that their structure be ordered by a relation of refinement. Given two G-structures X and Y, X is called a *refinement* of Y if and only if for each element in X there is a larger or equal element in Y; this is formally denoted as $X \overset{r}{\leq} Y$. The term 'larger' is used here strictly in the sense of the sets of variables associated with the elements. If X is a refinement of Y, then Y is called a *coarsening* of X. It is now established that the set of all G-structures for a given set of variables together with the refinement relation form a lattice referred to as a *refinement lattice* [24,72,75].‡ Given two G-structures, say X and Y, of a refinement lattice, X is called an immediate refinement of Y if and only if $X \overset{r}{\leq} Y$ and there is no Z such that $X \overset{r}{\leq} Z$ and $Z \overset{r}{\leq} Y$; Y is called an *immediate coarsening* of X. The set of all immediate refinements and all immediate coarsenings of a G-structure forms a complete *structure neighborhood* of the structure [24].§

‡Refinement lattices for three, four and five variables are specified in [75].

§All definitions introduced in this paragraph apply to G$^+$-structures as well.

It is obvious that the refinement ordering $\overset{r}{\leq}$ defined for G-structures with n variables is also a refinement ordering defined on the set of corresponding reconstruction hypotheses for any given overall system with n variables. By ordering reconstruction hypotheses via a refinement lattice, we are able to evaluate the various hypotheses by refinement level and by information distance. It has been proven that both probabilistic and possibilistic *information distances* [defined by equations (15) and (17), respectively], unlike other distance measures (e.g. Euclidean or Hamming), are *monotonically nondecreasing with increasing refinements.* Furthermore, these information distances are also *strictly additive* along any path in the refinement lattice [56]. Let $\overset{d}{\leq}$ denote the ordering of reconstruction hypotheses by the information distance. It is a connected weak ordering (reflexive, transitive and connected).

3.4 Solution Set

In the reconstruction problem, our aims are to maximize refinement and, at the same time, minimize information distance (loss). Unfortunately, these two aims conflict with each other. Any increase in the refinement implies that the distance increases or, at best, remains the same. When combined, the refinement and distance orderings ($\overset{r}{\leq}$ and $\overset{d}{\leq}$, respectively) form a *joint preference ordering.*

Given an overall system F and a set H_F of its reconstruction hypotheses, the joint preference ordering $\overset{*}{\leq}$ associated with the reconstruction problem is defined by

$$x \overset{*}{\leq} y \iff x \overset{r}{\leq} y \text{ and } x \overset{d}{\leq} y \text{ .}$$

for any pair $x, y \in H_F$. Following Gaines [48], the reconstruction problem can be then formulated as follows: given an overall system F and a set H_F of its reconstruction hypotheses (specified by the user), the *solution set* A_F of the reconstruction problem is a subset of H_F such that

$$A_F = \{x \in H_F \mid (\forall y \in H_F)(y \overset{*}{\leq} x \Rightarrow x \overset{*}{\leq} y)\}\text{ ,} \tag{22}$$

i.e. reconstruction hypotheses in A_F are either equivalent or incomparable in terms of the joint preference ordering $\overset{*}{\leq}$. It may be desirable to augment the two essential orderings with other (optional) preference orderings originating from various investigative contexts.

It is rather obvious that the whole refinement lattice or any desirable part of it can be searched by a repeated application of a simple procedure through which all immediate refinements are generated for any given reconstruction hypotheses. To describe such a procedure, it is sufficient to characterize each reconstruction hypothesis by the subsets of variables associated with its individual subsystems. Once a subset of variables is chosen to represent a subsystem, the constraint of these variables is uniquely determined from the constraint of the given overall system by calculating the appropriate projection according to the rules of the measure used in characterizing the constraints.

One procedure for generating all immediate refinements of a given reconstruction hypothesis is rather straightforward. Let a given reconstruction hypothesis be characterized by the family of subsets of variables $X = \{^k V \mid {}^k V \subset V, k \in K\}$, where V is the overall set of variables and K is a set of identifiers of the individual subsystem of the reconstruction hypothesis; clearly, subsystem k contains variables in set $^k V$. The following procedure generates all immediate refinements of X.

Procedure 1 [24,75]:

(i) Take one particular element of X that contains at least two variables, and replace it with all its subsets whose cardinalities are one less than the cardinality of the element itself.

(ii) Exclude any of the subsets introduced in Step (i) that are redundant; the result is an immediate refinement of X.

(iii) Repeat Steps (i) and (ii) for another element of X that contains at least two variables and has not been taken as yet; if there is no such an element of X, the procedure terminates.

It is obvious that this procedure can be extended to generalized reconstruction hypotheses by simply excluding the requirement [in Steps (i) and (iii)] that the element contains at least two variables.

A summary of basic issues involved in the reconstruction problem is expressed by Fig. 3.

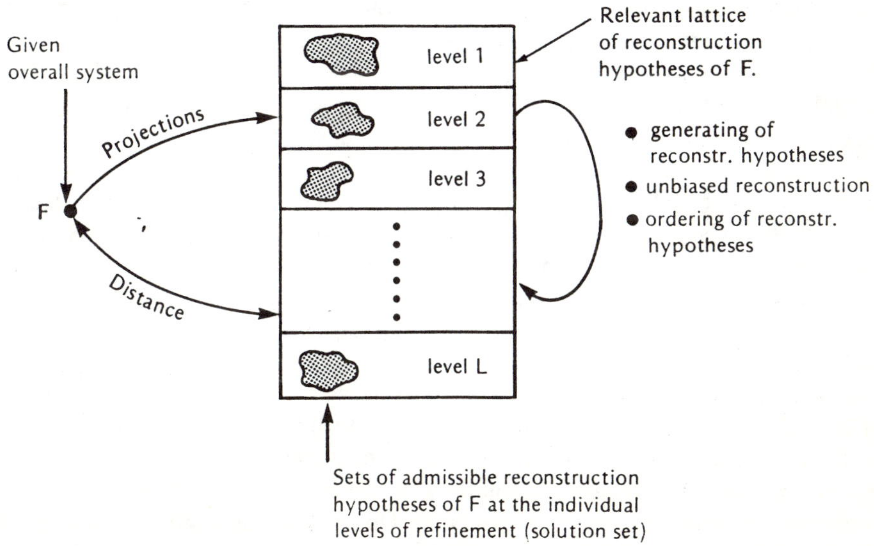

Figure 3 Summary of basic issues involved in the reconstruction problem.

3.5 Computational Levels

One of the main difficulties in dealing with the reconstruction problem is that the number of reconstruction hypotheses, each represented by one G-structure, increases extremely rapidly with the number of variables of the given overall system. The following table, where $|\mathcal{G}_n|$, $|\mathcal{G}_n^+|$ denote the cardinalities of the sets of all G-structures and G$^+$-structures for n variables, respectively, illustrates this point:

n	1	2	3	4	5	6	7
$\|\mathcal{G}_n\|$	1	2	9	114	6894	7,785,062	2,414,627,396,434
$\|\mathcal{G}_n^+\|$	1	4	18	166	7579	7,828,352	2,414,682,040,996

This table contains all currently known numbers for these sets.† The actual refinement lattices of G-structures for $n = 3,4,5$ are specified in [75].

To reduce the computational complexity involved in searching the refinement lattice, the lattice can be partitioned into *convenient equivalence classes* at several computational levels [24]. The equivalence classes can then be represented by appropriate *canonical G-structures* and special refinement or coarsening procedures developed to deal only with these canonical representations at the various computational levels. Three computational levels, referred to as *global, semi-global* and *local*, are described in previous papers [24,79].

Refinement lattices at the global level are homomorphic images of the corresponding lattices at the semi-global level, and the latter are, in turn, homomorphic images of the corresponding lattices at the local level [56]. Consequently, the notions of immediate refinements, immediate coarsenings, and structure neighborhood have different meanings at different computational levels [79].

To illustrate the difference between the computational levels, let us describe the main ideas associated with the global level. Given an overall system F defined by (1), each of its reconstruction hypotheses is uniquely determined by the associated G-structure, as argued in Section 3.1. For convenience, we may thus conduct the following discussion in terms of G-structures rather than reconstruction hypotheses.

Let each G-structure be characterized by an undirected graph, defined on the same set of variables V as the G-structure, which is constructed in the following way: for every pair of variables $v_i, v_j \in V$, v_i is connected with v_j in the graph if and only if both of these variables are contained (together) in at least one subset of the variables in the G-structure. This graph represents a symmetric and reflexive binary relation on V, which is usually called a compatibility or tolerance relation. Clearly, given a G-structure, the corresponding graph is unique. On the other hand, given a graph defined on the set V, there may be several G-structures that are characterized by it.

---

†$|\mathcal{G}_n^+|$ is equivalent to the number of monotone Boolean functions of n variables; it is Sequence No. 1439 in Sloane's *Handbook of Integer Sequences* (Academic Press, New York, 1973); $|\mathcal{G}_n|$ can easily be calculated from $|\mathcal{G}_n^+|$, $|\mathcal{G}_{n-1}^+|$,..., $|\mathcal{G}_1^+|$.

The mapping from the set of all G-structures defined on a set V into the set of all undirected graphs on V is thus many-to-one. As such, it imposes the following equivalence relation on the set of G-structures: G-structures G_i and G_j are *graph-equivalent* if and only if they are represented by the same graph; formally, we write $G_i \overset{g}{\equiv} G_j$. This equivalence relation partitions the set of all G-structures on V (and the corresponding reconstruction hypotheses) into equivalence classes in such a way that the refinement ordering is preserved [56]. Another advantage of the graph-equivalence is that the least refined and most refined G-structures in each of the equivalence classes are unique [24]. The least refined one, which is called a *C-structure*, consists of only and all maximal compatibility classes of the graph representing the equivalence class. The most refined one, which is called a *P-structure*, consists only of single variables or pairs of variables. As such, C-structures or P-structures can conveniently be used as alternative and complementary *canonical representations* of the equivalence classes. It is obvious that the number of undirected graphs that are reflexive is equal to

$$2^{\frac{n(n-1)}{2}},$$

where n is the number of variables involved (i.e. the number of nodes of the graphs); this is also the number of graph-equivalence classes of G-structures, as well as the number of C-structures and the number of P-structures.

The global procedure for generating reconstruction hypotheses consists of generating all immediate refinements (or coarsenings) of C-structures only (or, alternatively, of P-structures only), each of which represents an equivalence class of G-structures. The following is a refinement procedure for C-structures; it is assumed that a C-structure X and the associated graph $g(X)$ are given as an input to the procedure.

Procedure 2 [24,79]:

(i) Exclude one edge from the graph $g(X)$, say edge (v_i, v_j).

(ii) Split each element $x \in X$ that contains both v_i, v_j into two elements $x_i = x - \{v_j\}$ and $x_j = x - \{v_i\}$ and replace x by x_i and x_j.

(iii) Exclude all x_is and x_js generated in step (ii) which are subsets of some elements of X which were not replaced in step (ii) and record the result as an immediate C-structure refinement of X (together with the modified graph).

(iv) Repeat steps (i)-(iii) for all edges of the graph $r(X)$.

Since C-structures are the least refined (unique) G-structures in each equivalence class, all G-structures in each class can be generated by recursively refining its C-structure: refining it, then refining all its immediate refinements, etc. This can be done by a slightly modified Procedure 1: to stay in the same equivalence class, it must be required in Steps (i) and (iii) that the element contain at least three variables rather than two variables.

Several G-structures at the same level in the refinement lattice may have common immediate refinements. When a refinement procedure is applied to all of these G-structures, a number of duplicates may thus be generated. The problem of avoiding the generation of duplicates is discussed in [79].

In addition to procedures for generating immediate refinements, procedures for generating immediate coarsenings has been developed for the three computational levels as well [24,79]. When both refinement and coarsening procedures (at the same computational level) are applied to a G-structure, its complete immediate *structure neighborhood* is obtained—the set of all its immediate refinements and immediate coarsenings (at the respective computational level).

3.6 Reconstruction Problem Methodology

The reconstruction problem can be dealt with in a variety of alternative ways. It is useful, however, to view these alternatives solely as options of one underlying methodology, referred to as *reconstruction problem methodology* or *RP-methodology*.

A general scheme of the RP-methodology is expressed by the flow chart in Fig. 4. Regardless of the option employed, the methodology always consists of three kinds of procedures:

- procedures for generating relevant reconstruction hypotheses;
- procedures for evaluating the generated reconstruction hypotheses;
- procedures for deciding, after a set of generated reconstruction hypotheses has been evaluated, whether to stop or continue and, if it is decided to continue, how to generate additional reconstruction hypotheses.

In one way or another, procedures for generating reconstruction hypotheses are based upon the refinement ordering. They may differ from each other in at least three aspects. First, we distinguish refinement procedures from coarsening procedures. Second, both refinement and coarsening procedures may generate immediate successors in the refinement lattices (as Procedures 1 and 2 described in this paper do) or may skip some refinement levels according to the information obtained at the previous stage of the analysis. Third, the set of admissible reconstruction hypotheses may be restricted in some way. For example, instead of considering all G-structures, we may consider only C-structures, P-structures, loopless structures, or any other of the classes of structures shown in Fig. 5.† Furthermore, the user may restrict the class of reconstruction hypotheses by special requirements. For example, he may require that certain variables not be separated, that the reconstruction hypotheses do not contain more than some specific number of elements, that there be no shared variables between the elements of the reconstruction hypotheses, etc. The user may even request to generate only reconstruction hypotheses that form the structure neighborhood of a reconstruction hypothesis specified by him.

Procedures for evaluating reconstruction hypotheses depend predominantly upon the formal framework within which systems are conceptualized. Thus, for example, probabilistic systems require substantially different evaluation procedure than possibilistic systems. One way of evaluating a reconstruction hypothesis consists of determining its unbiased reconstruction (most conveniently by an appropriate join

†Classes of structures shown in Fig. 5 are defined in [24]. They are briefly characterized in the Appendix.

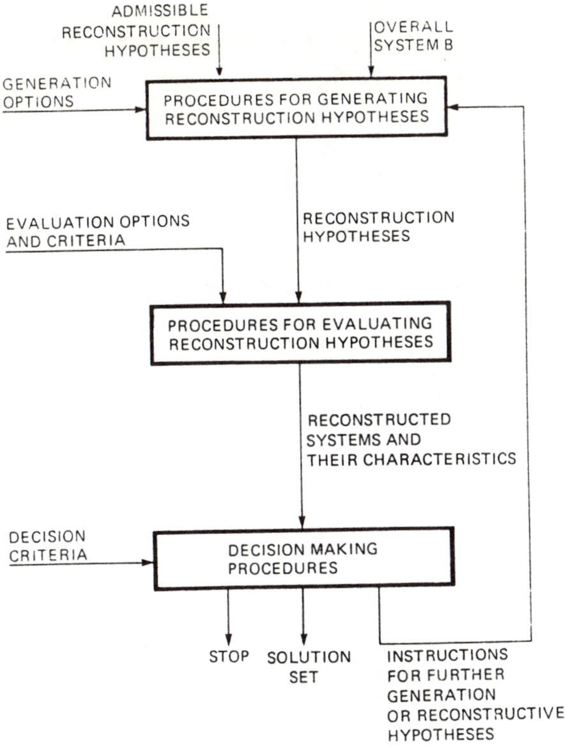

Figure 4 General scheme of the RP-procedure.

procedure) and calculating the information distance (or some other distance such as Hamming or Euclidean) between the unbiased reconstruction and the given overall system [27,30,75]. One advantage of this evaluation strategy is that it can be easily modified when one formal framework is replaced with another one. For example, when we want to switch from probabilistic systems to possibilistic systems, we need to replace only the formula representing the join operation and the formula of the information distance; everything else remains the same. For probabilistic systems, a number of special strategies for evaluating reconstruction hypotheses, which are based on various aspects of Shannon's information theory, have been developed by Broekstra [11-20], Conant [33-44], Dussauchoy [46] and Krippendorff [80-85]. In addition, the evaluation of probabilistic reconstruction hypotheses has also been studied without the use of information theory. One of the most comprehensive presentations of the various methods in this category is given in the book by Bishop *et al.* [10].

The purpose of the decision-making procedures of the RP-methodology (Fig. 4) is to control the search through the set of admissible reconstruction hypotheses, to select members of the solution set, and to determine when the overall analysis should terminate. These procedures are conditioned by decision criteria specified by the user. For example, the user may require that the analysis terminate at that level of

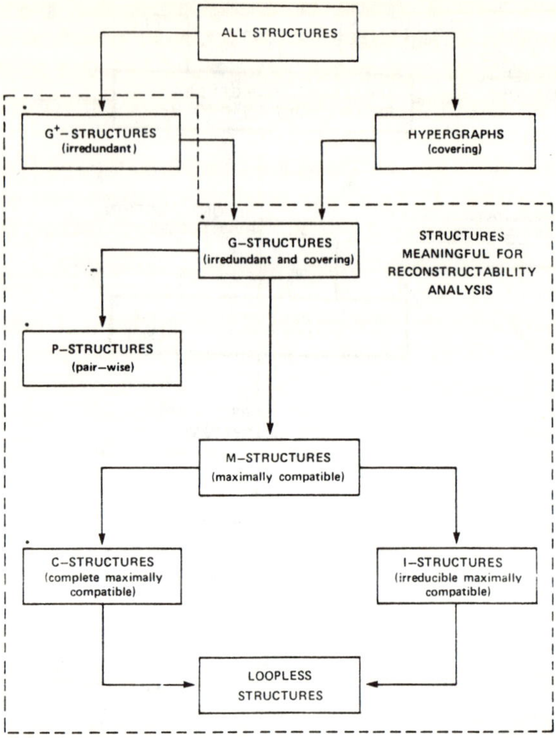

Figure 5 Classification of structures.

refinement where all distances are greater than a specified value or he may require that at each level of refinement only structures with minimal distance at that level are further refined, etc. [29].

Examples of the use of the RP-methodology can be found in numerous articles (e.g. [27,55,70,71,78,98]) and a book [75].

4 METAMETHODOLOGICAL ASPECTS OF RECONSTRUCTABILITY ANALYSIS

Extensive simulation experiments have been performed since 1976 to determine some fundamental characteristics of RA, particularly the various options of the RP-methodology. The purpose of these experiments is three-fold: (i) to get a deeper understanding of RA; (ii) to help systems investigators in making the various decisions in the process of systems modeling, and (iii) to test some new hypothetical principles associated with RA such as a new principle of inductive inference proposed by Klir in 1981 [73] and described later in this paper.

In a typical experiment regarding a particular option of the RP-methodology, a reconstruction hypothesis is selected for a given number of variables and their resolutions (i.e. their associated state set cardinalities). By performing the appropriate join procedure, the true overall system is derived from this given reconstruction hypothesis. The process of generating data by the *given hypothesis* is then simulated by computer. In most of the experiments, sequences of 2000 data points are generated. RA is performed on 10 different segments of each of these data sequences and the results are compared with the *true overall systems*. For each data segment the frequencies of the individual states are calculated and then converted to probabilities or possibilities, respectively. These probability or possibility distributions represent an overall system that is called a *data-bound system*. This system is derived from all observations present in the data by considering all variables simultaneously. At this point, the investigated option of the RP-methodology is performed on the data-bound system. For the sake of simplicity, the experiments were restricted to C-structures. Specific attention is given to the level of refinement corresponding to that of the true structure system. If the best hypothesis at that level is the same as the true structure system, the experiment is considered a success. In any case, however, the elements of the hypothesis that correspond to the true structure system are joined to form the unbiased reconstruction called the *reconstructed overall system*. This system and the data-bound system are then compared with the original (true) overall system.

By performing over 12,000 experiments of this sort, a database was formed relating reconstruction properties (e.g. various distances and performance measures) with various parameters such as the number of variables, their resolutions, the number of observations and refinements levels [52]. Various characteristics were then determined by summarizing appropriate sets of data from this database. Some of the characteristics confirmed the proposed hypothetical inductive principle [73], according to which the reconstructed overall system is closer to the corresponding true overall system than its data-bound counterpart.

The experiments are described in more detail in a recent paper [54]. All details as well as summarized results are available in an archive referred to as *General Systems Depository* [53].

The described experimental results characterize some aspects of the RP-methodology. This means that they are metamethodological results. Such results are particularly important for developing expert systems in various areas of systems science. These expert systems, exemplified by the *General Systems Problem Solver* [75], will have to contain not only knowledge and methodological bases, but an adequate metamethodological base as well. The latter should contain characteristics of the various methodologies available in the expert system (such as induction characteristics of the RP-methodology), and the associated guidelines for the user.

5 FURTHER RESEARCH ON RECONSTRUCTABILITY ANALYSIS

At least three issues associated with the reconstruction family require further research. First, the formulation of a meaningful *identifiability quotient* for probabilistic systems. Second, the development of user-convenient forms for representing reconstruction families. Third, the investigation of a *least risk (min-max) reconstruction*, from both procedural and interpretive points of view. Such a reconstruction minimizes the largest information distance that can occur between it and the other members of the reconstruction family, i.e. it should represent the 'midpoint' of the reconstruction family. This investigation will require the use of general information distances defined by formulas (18) and (20). While the possibilistic version is fully satisfactory since it is a metric distance, the probabilistic counterpart does not possess the triangle property of metric distances [91]. Further research is thus needed to derive a more adequate (i.e. metric) information distance for probabilistic systems.

All previous research on RA has been based on the assumption that we deal with structure systems that are consistent. In the reconstruction problem, where elements of each reconstruction hypothesis are derived from the given overall system as projections, the consistency (both local and global) is guaranteed. In the identification problem, however, where functions ${}^kf$ in a structure system defined by (2) are derived from data sets collected independently of each other, possibly by different experimental teams, the resulting structure system may be locally inconsistent. Such an inconsistency (in contrast with a global inconsistency) does not imply that the variables investigated are themselves inconsistent. It simply reflects the fact that we have only incomplete information regarding each of the subsets of variables involved in the structure system. It is this incompleteness that creates the local inconsistency and, consequently, it is meaningful and desirable to accept such structure systems and deal with them.

Locally inconsistent structure systems are obviously resistant to the use of regular logical procedures. They require that we first apply some procedures through which their internal inconsistencies are resolved. Interest in such procedures has been shown only recently [87] and further research is needed. We view this problem basically as the following optimization problem: given a locally inconsistent structure system of form (2), modify functions ${}^kf\,(k \in N_q)$ in such a way that the new functions, say functions ${}^kf'$, are locally consistent and as close to the original ones as possible in terms of the appropriate information distance (or, possibly, in some other specific sense). It was desirable at the initial stage of research on RA to ignore the various *distinctions among the variables* involved, such as directly observed variables, lagged or internal (postulated) variables, input vs. output variables, etc. Although these distinctions among variables do not greatly affect the procedures of RA, they are important for properly interpreting the results of RA. For example, they are essential for the characterization of dynamical properties of systems. They also introduce additional requirements for local consistency of the structure system involved. A comprehensive study of all issues related to the various distinctions among variables in the context of RA is required to understand the role of RA in dealing with dynamical systems.

Thus far, RA has been developed only for probabilistic and possibilistic systems. Further research is needed to extend it to systems conceptualized within other types of fuzzy measures. Plausibility theory, which subsumes both probability and possibility theory, is of particular significance [92].

The main difficulty in dealing with the reconstruction problem is the search through the refinement lattice. Computational demands grow so rapidly that searching through the whole lattice (e.g. by Procedure 1) is possible only for a small number of variables, say six or less. The most fundamental aim is to develop procedures that would require a search only through a small portion of the refinement lattice and, yet, be able to determine the reconstruction hypotheses with the minimum distance at each level of refinement. This will require a thorough theoretical investigation of the refinement lattices from the standpoint of the information distance in order to derive useful lower and upper bounds of dependencies between values of information distance at the various refinement levels. Such an investigation is feasible since the information is strictly additive along any refinement path in the lattice.

It is also known that the reconstruction family is monotonically nondecreasing (in the set theoretic sense) with increasing refinements [56]. However, this important property has not been utilized in the RP-methodology as yet.

It is desirable to determine theoretical and experimental estimates of computational complexities for the various options of the RP-methodology as well as the methodology for the identification problem. Related to this issue are some unsolved combinatorial problems such as the number of G-structure, maximum number of elements in G-structures, maximum number of cliques in reflexive undirected graphs, number of loopless G-structures or C-structures, etc.

As previously mentioned, the various *characteristics of the various options of the RP-methodology obtained by simulation experiments* are extremely important to help the users of RA to properly interpret their results. All previously performed experiments (as described in Section 4) were based on the assumption that the data were generated by a specific reconstruction hypothesis. Their objective was to determine how scarcity of data affects RA. Although such idealized experimentation is valuable and represents a natural first stage in the evaluation of RA, it is highly desirable to extend it to more general situations. In a generalized experiment, the data should be generated from an arbitrary overall system, which is not necessarily reconstructable from a structure system. Its reconstruction properties—the complete list of relevant reconstruction hypotheses, ordered at each refinement level by their information distances—has to be determined and, then compared with the corresponding reconstruction properties obtained for the systems derived from the various segments of the generated data.

The mentioned reconstruction principle of inductive reasoning has been partially confirmed by the simulation experiments described in Section 4. Its ultimate confirmation will require the generalized experiments outlined in the previous paragraph. The inductive principle consists of three steps: (i) superior reconstruction hypotheses are determined for the overall system at the various refinement levels [this is basically the solution set defined by equation (22)]; (ii) beliefs of various degrees that these superior hypotheses reflect the actual reconstruction properties of the variables involved are formed on the basis of relevant experimental characteristics, resulting in general guidelines or a specific belief function; (iii) the given overall constraint is supplemented with (or replaced by) the constraints reconstructed by the superior reconstruction hypotheses, each associated with its respective degree of belief.

Research on the reconstruction principle of inductive inference is needed in two main thrusts: (i) a study of the reconstruction characteristics obtained by the idealized simulation experiments (as described in Section 4) as well as the proposed generalized experiments for the purpose of developing meaningful guidelines for determining the degree of belief in the information gained by the principle under various conditions (number of variables, their resolutions, data size, etc.); and (ii) a theoretical investigation of the principle that should result in its better understanding and justification.

There are of course many additional issues associated with RA that require further research. The research issues outlined in this section are some of the most important ones for making RA an attractive and powerful methodological tool.

REFERENCES

Abbreviations of two frequently used periodicals in this bibliography:

IJGS—International Journal of General Systems
SGSR—Proceedings of the Annual Meeting of the Society for General Systems Research.

1. W. R. Ashby, Constraint analysis of many-dimensional relations. *Gen. Syst. Yearbook* **9** (1964), 99-105.
2. W. R. Ashby, Constraint analysis of many-dimensional relations. In N. Wiener and J. P. Schade (eds.), *Progress in Biocybernetics*, pp. 10-18. Elsevier, Amsterdam (1965).
3. W. R. Ashby, Measuring the internal informational exchange in a system. *Cybernetica* **1** (1965), 5-22.
4. W. R. Ashby, Two tables of identities governing information flows within large systems. *ASC Commun.* **1** (1969), 3-8.
5. W. R. Ashby, Information flows within co-ordinated systems. In J. Rose (ed.), *Progress in Cybernetics*, Vol. 1, pp. 57-64. Gordon & Breach, London (1970).
6. W. R. Ashby, Systems and their informational measures. In G. J. Klir (ed.), *Trends in General Systems Theory*, pp. 78-97. Wiley-Interscience, New York (1972).
7. Y. Bard, Estimation of state probabilities using the maximum entropy principle, *IBM J. Res. Dev.* **24** (1980), 563-569.
8. Y. Bard, A model of shared DASD and multipathing. *ACM Commun.* **23** (1980), 564-572.
9. C. Berge, *Graphs and Hypergraphs*. North-Holland/American Elsevier, Amsterdam and New York (1973).
10. Y. M. M. Bishop, S. E. Fienberg and P. W. Holland, *Discrete Multivariate Analysis*. MIT Press, Cambridge, Mass. (1975).
11. G. Broekstra, Constraint analysis and structure identification. *Annls Syst. Res.* **I** (1976), 67-80; **II** (1977), 1-20.
12. G. Broekstra, On the representation and identification of structure systems. *IJGS* **9** (1978), 1271-1293.
13. G. Broekstra, Two methods of constraint analysis. In *Progress in Cybernetics and Systems Research*, pp. 13-23. Hemisphere, Washington, D.C. (1980).
14. G. Broekstra, Structure modeling: a constraint (information) analytic approach. In G. J. Klir (ed.), *Applied General Systems Research*, pp. 117-132. Plenum Press, New York (1978).
15. G. Broekstra, Probabilistic constraint analysis for structure identification: an overview and some social science applications. In B. P. Zeigler (ed.), *Methodology in Systems Modeling and Simulation*, pp. 305-334. North-Holland, Amsterdam (1979).
16. G. Broekstra, (Non) probabilistic constraint analysis and two-stage approximation method of structure identification. *SGSR* (1979), 73-81.
17. G. Broekstra, C-analysis: the application of informational measures to the reconstructability problem. *SGSR* (1980), 2-8.
18. G. Broekstra, C-analysis of C-structures. *IJGS* **7** (1981), 33-61.
19. G. Broekstra, Structural relation models: improved quality of qualitative modeling. In G. E. Lasker (ed.), *Applied Systems and Cybernetics*, pp. 953-957. Pergamon Press, New York (1981).
20. G. Broekstra, Constraint analysis of discrete multivariate systems. *SGSR* (1982), 259-265.
21. D. T. Brown, A note on approximations to discrete probability distributions. *Inform. Control* **2** (1959), 386-392.

22. R. E. Cavallo, *The Role of Systems Methodology in Social Science Research.* Martinus Nijhoff, Boston (1979).
23. R. E. Cavallo, Reconstructability and identifiability in the evaluation of structure hypotheses: an issue in the logic of modeling, *SGSR* (1980), 647-654.
24. R. E. Cavallo and G. J. Klir, Reconstructability analysis of multi-dimensional relations: a theoretical basis for computer-aided determination of acceptable systems models. *IJGS* **5** (1979), 143-171.
25. R. E. Cavallo and G. J. Klir, The structure of reconstructable relations: a comprehensive study. *J. Cybernet,* **9** (1979), 399-413.
26. R. E. Cavallo and G. J. Klir, Reconstructability analysis: overview and bibliography. *IJGS* **7** (1981), 1-6.
27. R. E. Cavallo and G. J. Klir, Reconstructability analysis: evaluation of reconstruction hypotheses. *IJGS* **7** (1981), 7-32.
28. R. E. Cavallo and G. J. Klir, Systems of overall possibilities: reconstruction aspects. In G. E. Lasker (ed.), *Applied Systems and Cybernetics*, pp. 939-946. Pergamon Press, New York (1981).
29. R. E. Cavallo and G. J. Klir, Decision making in reconstructability analysis. *IJGS* **8** (1982), 243-255.
30. R. E. Cavallo and G. J. Klir, Reconstruction of possibilistic behavior systems. *Fuzzy Sets Syst.* **8** (1982), 175-197.
31. R. Christensen, *Entropy Minimax Sourcebook, Vol. I: General Description.* Entropy, Lincoln, Mass. (1981).
32. F. L. Comstock and H. J. J. Uyttenhove, A systems approach to grading of flight simulator students. *J. Aircraft* **16** (1979), 780-786.
33. R. C. Conant, Detecting subsystems of a complex system. *IEEE Trans. Syst., Man and Cybernet.* **SMC-2** (1972), 550-553.
34. R. C. Conant, Measuring the strength of intervariable relations. *Kybernetes* **2** (1973), 47-52.
35. R. C. Conant, Two comments on structural modeling. *Proc. 1979 IEEE Int. Conf. Cybernet. Soc.*, Denver, pp. 404-408 (1979).
36. R. C. Conant, Structural modeling using a simple information measure. *Int. J. Syst. Sci* **11** (1980), 721-730.
37. R. C. Conant, A structural modeling technique using information measures. *SGSR* (1980), 9-18.
38. R. C. Conant, A two-stage technique for structure modeling. *Proc. 1980 IEEE Int. Conf. Cybernet. Soc.* Cambridge, Mass., pp. 465-467 (1980).
39. R. C. Conant, Detection and analysis of dependency structures. *IJGS*7 (1981), 81-91.
40. R. C. Conant, Set-theoretic structure modeling. *IJGS*7 (1981), 93-107.
41. R. C. Conant, Approximately general structures with dependent analysis. *SGSR* (1982), 243-249.
42. R. C. Conant, Combinatorial constraints on structure analysis. In G. E. Lasker (ed.). *Applied Systems and Cybernetics*, pp. 947-952. Pergamon Press, New York (1981).
43. R. C. Conant, Identifying relations from partial information. *SGSR* (1981), 463-468.
44. R. C. Conant, Structure analysis of large systems using heuristics. *SGSR* (1984), 231-235.
45. D. Dubois and H. Prade, *Fuzzy Sets and Systems*. Academic Press, New York (1980).
46. R. L. Dussauchoy, Generalized information theory and decomposability systems. *IJGS* **9** (1982), 13-36.
47. D. Elias, Procedures for generating reconstruction hypotheses in the reconstructability analysis: APL version. *General System Depository* No. 072179 (IFSR, Hofstrasse 1, A-2361 Laxenburg, Austria) (1972).
48. B. R. Gaines, System identification, approximation and complexity, *IJGS* **3** (1977), 145-174.
49. J. A. Goguen and F. J. Varela, Systems and distributions: duality and complementarity. *IJGS* **5** (1979), 31-43.
50. A. R. Gourlay, J. M. McLean and P. Shepherd, Identification and analysis of the subsystem structure of models. *Appl. Math. Model.* **1** (1977), 245-252.
51. S. Guiasu, *Information Theory with Applications*. McGraw-Hill, New York (1977).
52. A. Hai, An empirical investigation of reconstructability analysis. Ph.D. dissertation, Thomas J. Watson School of Engng, Applied Science, and Technology, SUNY-Binghamton, New York (1984).

53. A. Hai and G. J. Klir, Reconstructability analysis: detailed and summarized experimental data. *General Systems Depository* (IFSR, Hofstrasse 1, A-2361 Laxenburg, Austria) (1984).
54. A. Hai and G. J. Klir, An empirical investigation of reconstructability analysis. *Int. J. Man-Machine Studies* (forthcoming).
55. M. Higashi, Applicability of reconstructability analysis to ecology. *SGSR* (1982), 1056-1061.
56. M. Higashi, A systems modeling methodology: probabilistic and possibilistic approaches, Ph.D. dissertation. T. J. Watson School of Engng, SUNY-Binghamton, New York (1983).
57. M. Higashi and G. J. Klir, Measures of uncertainty and information based on possibility distributions. *IJGS* **9** (1983), 43-58.
58. M. Higashi and G. J. Klir, On the notion of distance representing information closeness: possibility and probability distributions. *IJGS* **9** (1983), 103-115.
59. M. Higashi and G. J. Klir, Resolution of finite fuzzy relation equations. *Fuzzy Sets Syst.* **13** (1984), 65-82.
60. M. Higashi, G. J. Klir and M. A. Pittarelli, Reconstruction families of possibilistic structure systems. *Fuzzy Sets Syst.* **12** (1984), 36-60.
61. E. T. Jaynes, Prior probabilities. *IEEE Trans. Syst. Sci. Cybernet.* **SSC-4** (1968), 227-241.
62. E. T. Jaynes, Where do we stand on maximum entropy? In R. L. Levine and M. Tribus (eds.), *The Maximum Entropy Formalism*, pp. 15-118. MIT Press, Cambridge, Mass. (1979).
63. B. Jones, Determination of reconstruction families. *IJGS* **8** (1982), 225-228.
64. B. Jones, Determination of unbiased reconstructions. *IJGS* **10** (1984).
65. B. Jones, Recent algorithms in reconstructability analysis. *SGSR* (1984), 243-245.
66. G. A. Kengerlinskiy, An informational approach to the decomposition of complex systems. *Engng Cybernet.* **16** (1978), 91-97.
67. G. J. Klir, Identification of generative structures in empirical data. *IJGS* **3** (1976), 89-104.
68. G. J. Klir, The problem of choosing appropriate structure system in systems modeling. *Proc. Int. Conf. Syst. Modeling in Developing Countries,* Asian Inst. Technol., Bangkok, pp. 3-20 (1978).
69. G. J. Klir, Structural modeling of indigenous systems. *SGSR* (1978), 151-155.
70. G. J. Klir, General systems problem solving methodology. In B. Zeigler (ed.), *Modeling and Simulation Methodology*, pp. 3-28, North-Holland, Amsterdam (1979).
71. G. J. Klir, Computer-aided systems modeling. In E. Halfon (ed.), *Theoretical Systems Ecology*, pp. 291-323. Academic Press, New York (1979).
72. G. J. Klir, Architecture of structure systems: a basis for the reconstructability analysis. *Acta Polytechnica Scandinavica*, Mathematics and Computer Science Series, No. 31, Helsinki, pp. 33-43 (1979).
73. G. J. Klir, On systems methodology and inductive reasoning: the issue of parts and wholes. *Gen. Syst. Yearbook* **26** (1981), 29-38.
74. G. J. Klir, Reconstructability analysis: an overview. In T. Oren (ed.), *Simulation and Model-Based Methodologies*. Springer-Verlag, New York (1984).
75. G. J. Klir, *Architecture of Systems Problem Solving*. Plenum Press, New York (1985).
76. G. J. Klir and A. Hai, The computer as a meta-methodological tool: a case of reconstructability analysis. *SGSR* (1984), 237-242.
77. G. J. Klir and H. J. J. Uyttenhove, Computerized methodology for structure modeling. *Ann. Syst. Res.* **5** (1976), 29-66.
78. G. J. Klir and H. J. J. Uyttenhove, On the problem of computer-aided structure identification: some experimental observations and resulting guidelines. *Int. J. Man-Machine Studies* **9** (1977), 593-628.
79. G. J. Klir and H. J. J. Uyttenhove, Procedures for generating reconstruction hypotheses in the reconstructability analysis. *IJGS* **5** (1979), 231-246.
80. K. Krippendorff, A spectral analysis of relations: further developments. *Proc. Fourth European Meeting on Cybernetic and Systems Research*, Linz, Austria (Hemisphere, Washington, D.C.) (1978).
81. K. Krippendorff, On the identification of structures in multivariate data by the spectral analysis of relations. *SGSR* (1979), 82-91.
82. K. Krippendorff, On the identification of latent functions in multivariate systems. *Proc. Fifth European Meeting on Cybernetics and Systems Research,* Vienna (Hemisphere, Washington, D.C.) (1980).

83. K. Krippendorff, An algorithm for identifying structural models of multi-variate data. *IJGS* **7** (1981), 63-79.
84. K. Krippendorff, A proposal for an algorithm for generating loopless or recursive models of multivariate data. *SGSR* (1982), 299-304.
85. K. Krippendorff, Regression analysis using information theory. *SGSR* (1982), 1007-1012.
86. R. F. Madden and W. R. Ashby, On the identification of many-dimensional relations. *Int. J. Syst. Sci.* **3** (1972), 343-356.
87. M. Mariano, Toward resolving inconsistency among behavior systems. *SGSR* (1984), 225-229.
88. J. H. Marschal, The concept of a system. *Phil. Sci.* **42** (1975), 448-467.
89. M. Pittarelli, Computational and semantic aspects of possibilistic constraint systems. *SGSR* **1** (1984), 219-224.
90. Reconstructability analysis of Iragi soils and crops by a systems modeling approach. Report, University of Ghent, Belgium (1983).
91. P. K. Sahoo, On Higashi and Klir's conjecture on a measure of information closeness. *IJGS* **10** (1984), 1-4.
92. G. Shafer, *A Mathematical Theory of Evidence*. Princeton University Press, Princeton (1976).
93. J. E. Shore and R. W. Johnson, Axiomatic derivation of the principle of maximum entropy and the principle of minimum cross-entropy. *IEEE Trans. Informant. Theory* **IT-26** (1980), 26-37.
94. Special Issue of the *Int. J. Gen. Syst. Reconstructability Anal.* **7** (1981), 1-107 (Broekstra, Cavallo, Conant, Klir, Krippendorff).
95. P. Suppes, Reductionism, atomism and holism. In *Parts & Wholes*, Documents from Int. Workshop, Stockholm, pp. 134-140 (unrevised edition) (1983).
96. H. J. J. Uyttenhove, Computer-aided systems modeling: an assemblage of methodological tools for systems problem solving. Ph.D. dissertation, School of Advanced Technology, SUNY-Binghamton, New York (1978).
97. H. J. J. Uyttenhove, *SAPS (Systems Approach Problem Solver): An Introduction and Guide*. Computing and Systems Consultants, Binghamton, New York (1981).
98. H. J. J. Uyttenhove, Systems approach problem solver and the open heart surgery patient. In R. Trappl (ed.), *Cybernetics and Systems Research*, pp. 655-661, North-Holland, Amsterdam (1982).
99. H. J. J. Uyttenhove and P. Gomez, The systems approach problem solver and a stockmarket example. *SGSR* (1981), 341-348.
100. R. E. Valdes-Perez and R. C. Conant, Information loss due to data quantization in reconstructability analysis. *IJGS* **9** (1983), 235-247.
101. J. C. Wanser, System modeling of an archaeological site. General systems workshop project, Department of Systems Science, SUNY-Binghamton, New York (1980).
102. L. A. Zadeh, Fuzzy sets as a basis for a theory of possibility. *Fuzzy Sets Syst.* **1** (1978), 3-28.

APPENDIX: CLASSES OF STRUCTURES RELEVANT TO RA

The term '*structure*' is used in RA for families of subsets of a given set of variables, say set V. There are 2^{2^n} structures of this sort for n variables. Various special classes of structures are illustrated by the diagram in Fig. 5. The arrows in the diagram indicate a subset relationship between the classes.

A special class of structures, which are relevant in some instances to RA, are structures that satisfy the irredundancy requirement (14). They are called in this paper *extended general structures* or *G^+-structures*.

Another special class of structures are known as *hypergraphs*. A hypergraph is a family of subsets of V that covers all elements of V [i.e. it satisfies the covering condition (13)] and does not contain the empty set. A special class of hypergraphs are *G-structures*, which represent reconstruction hypotheses in RA. They satisfy the covering condition (13) and irredundancy condition (14).

The five subclasses of G-structures shown in Fig. 5 are based on undirected and reflexive graphs (symmetric and reflexive relations) defined on V. P-structures and C-structures are introduced in Section 3.5. A *P-structure* consists of all pairs (v_i, v_j) of different variables $(i \neq j)$ that correspond to edges in the associated graph and all single variables that are not redundant. A *C-structure* consists of only and all maximal compatibility classes (cliques) of the associated graph. Hence, there is one P-structure and one C-structure for each graph. For n variables, there are exactly $2^{\frac{n(n-1)}{2}}$ of either of them.

M-structures also consist only of maximal compatibility classes of the associated graphs, but are not required to contain all of the classes (contrary to C-structures). Hence, the class of M-structures is larger than that of C-structures. *I-structures* consist of irreducible coverings of the associated graphs by maximal compatibility classes. There is a partial overlap between the classes of C-structures and I-structures, but neither of them is a subset of the other.

Loopless structures are defined as structures that do not contain any loop of the following kind: let

$$E_{k_1}, E_{k_2}, E_{k_3}, \ldots E_{k_m}$$

be a sequence of elements of a given structure (sequence of subsets of V) such that

$$E_{k_1} \cap E_{k_{i+1}} \neq \varnothing$$

and $m \geq 3$; then, this sequence represents a loop if and only if there exists a variable $v_x \in V$ such that $v_x \in E_{k_1}$, $v_x \in E_{k_m}$, and $v_x \notin E_{k_i}$ for $i \neq 1, m$. Loopless structures form a special class of C-structures as well as I-structures.

DISCUSSION OF KLIR'S PRESENTATION

Kossov

Regarding inconsistency of analysis:

1) May one consider that in certain cases the paradoxes we talked about yesterday arise because of the inconsistency of the analysis or are they inherent in the subsystem to which the paradox belongs? Would they disappear if a more complete system was used?

2) You showed examples (pictures) in which the parts (subsystems) seemed consistent but the whole was not. Are there any examples of the opposite?

Klir

1) They are inherent in the subsytems. Further investigation of the subsystems may resolve the local inconsistencies.

2) Global inconsistency is defined as an inconsistency in the structure of a system that occurs even when all subsystems are locally consistent. Hence, I believe that, by definition, the opposite case is logically not possible.

Pospelov

Is there a relationship between the number of elements and the number of constraints assumed? If two subsystems have common arguments, how can they be divided or separated?

Klir

In reconstructability analysis we deal with collections of subsystems of a given overall system that are not necessarily disjoint. Each subsystem is associated with a subset of variables of the full set of variables under consideration. We thus deal with hypergraphs defined on the given set of variables rather than partitions of the set. The variables that are shared by two subsystems represent the interaction between them.

Reckmeyer

Paradoxes exist because of our inability to *consciously* address our world in an entirely consistent fashion, so expanding the scope of the analysis will not resolve all paradox—only some, for there are always epistemological, ontological, and methodological aspects to paradoxes. The fundamental concern here is how to identify and deal with the salient few when dealing with so many variables?

Klir

I assume that the term "paradox" means in this discussion some sort of inconsistency. In my view, there are two kinds of inconsistencies—local and global. Local inconsistencies can be, in principle, resolved. A general treatment of this area is included, e.g., in the book, *Logic of Inconsistency*, by Nicolas Rescher. Global inconsistency is more serious and indicates that our system is ill-conceived.

Lapin

In economic planning there is the problem of analyzing reserved land for future economic use. Inasmuch as the land is located within broader regional systems and it is not known beforehand what type of plant will be constructed there, 1) can reconstructability analysis be used to solve the problem of how such land can be used? And 2) has any work already been done on this problem?

Klir

Reconstructability analysis has not been used in the problem proposed. It is an interesting problem, and I believe that reconstructability analysis could, in principle, be applied to the problem even though some modification of the current methodology would probably be necessary.

Mitroff

Sometimes it is necessary to destroy the system in order to understand it. There is a moral constraint on learning about some systems—the most fundamental ones. Global problems are even more complex. Paradoxes will not vanish by taking a systems approach, but the systems approach will reveal new paradoxes.

Klir

It all depends on what we mean by the term "paradox." I do not argue that all paradoxes can be eliminated by our understanding of the part-whole relationship, but I believe that some of our paradoxes (in the broad sense of the term, as used in our discussions yesterday) are a result of our poor understanding of this relationship, i.e., we often make wrong inferences from incomplete information.

Rapoport

If your methodology is proper for the study of social systems (as you said), how do you handle the kind of subsystem which has variable properties and whose changes are not predictable? Social systems include many such subsystems and as far as I know they define the behavior of the whole.

Klir

Such subsystems exist, but we are not considering their behavior and their properties at present. That would be a more general case, which is not the subject-matter of our consideration here.

Kossov

It seems that a consistent analysis of a more general system, extending beyond the framework of an observed paradox, can yield a key to resolving this paradox. In this case new alternatives may appear which have not been observed within the framework of the subsystem.

Klir

There are various reasons for an inconsistency in the structure of a system. Local inconsistencies, in general, are a result of our ignorance regarding the subsystems. Hence we can resolve them by modifying the subsystems in such a way that the structure becomes consistent, and yet, the loss of information in the subsystems is minimized. Global inconsistency is more serious. It means that the structure of the system is ill-conceived, and it is questionable whether we should even try to resolve it.

Sadovsky

To what kind of social system do you apply your methodology of reconstructability analysis? And what are the results?

Klir

Some examples:

1) A study of several thousand government employees, which included variables such as salary, education, length of employment, sex, age, etc., with the objective of identifying possible inequities. The study did discover some inequities

2) Roger Cavallo describes an application of reconstructability analysis in his book, *Systems Methodology in Social Science Research.* It is a study of violence in about 50 different countries.

3) I have studied the impact of some political variables on the performance of the U.S. stock market.

CHAPTER

8

LOGIC IN ARTIFICIAL INTELLIGENCE SYSTEMS

Dmitry A. Pospelov

Logic models are pivotal to artificial intelligence. They are used in knowledge acquisition, generation of new notions, decision making, planning of purposeful robot behavior, and in many other case.

Artificial intelligence, however, not only employs the traditional logic models such as syllogistics or predicate calculus, but also contributes significantly to the origination of new logical systems whose properties are defined by the artificial intelligence problems.

First of all, the notion is presented of a formal system which is basic to the classical logic models. Formal system is defined in the form of <T, SR, A, IR> where T is a set of basic system elements; SR is a set of syntactic rules enabling determination of all the correct syntactic constructs from the elements of T; A is a set of axioms; and IR is a set of inference rules. The insufficiency of formal systems manifests itself even when using logic in knowledge representation systems. It becomes necessary to consider instead semiotic systems defined as <F, CH(T) CH(BR) CH(A) CH(IR)> where F is a formal system, and CH are rules for transformation of corresponding elements of the formal system. These transformations reflect dynamics of the content of AI system databases. In this case, usually the elements from T have their own structure and characteristics (e.g., are frames or products) and the rules from P generate from them heterarchical network structures: semantic networks, frame networks, etc. This fact allows one to assert that logics on relations or their equivalent algebraic relational systems are the basic form of logic models in AI.

The rules CH(A) introduce changes of "possible worlds" which the AI system deals with [1], and relations reflecting situational variability of (temporal, spatial, etc.) facts and events require that logic system dynamics be taken into account [2]. This have given rise to the logic systems that were studied by the turn of the sixties in situational control systems [3,4]. Many ideas used in the models were later elaborated in AI. Anyway, they have proved sufficient for problems concerned with knowledge representation and manipulation.

The rules CH(I) enable the artificial systems to simulate adaptation to a changing environment, and CH(T) and CH(SR) allow one to introduce into knowledge bases new notions and syntactically correct constructs, i.e. to learn new means for environmental descriptions.

Simulation of human reasoning in intellectual systems (expert systems, robots, query-response systems, etc.) is especially important for AI logic systems. There is a bulky literature on this subject. Let us list some basic requirements of AI systems for logic models.

A. In any traditional logic system, it follows from the possibility to infer facts ϕ_1 and ϕ_2 that the conjunction ϕ_1 and ϕ_2 may be inferred as well. It may be readily demonstrated that this requirement is not obeyed when reasoning about future events. Consider an example of a robot standing at the point (x,y) and planning its actions (Fig. 1). Its logic system infers two independent facts: ϕ_1—it is possible to move ahead to the point (x',y'), and ϕ_2—the manipulator may be rotated quarter turn to the left. However, if the fact ϕ_1 and ϕ_2 is regarded as true, the manipulator will be inevitably broken against the tree standing on the way from (x,y) to (x',y').

Here, the situativeness of reasonings makes itself manifest. Being true in one situation (in the above example, when the robot is at the point (x,y)), they become false in other. In AI systems, the situativeness of reasoning usually is explicitly taken into account in the form of special conditions imposed on decision making and allowing one to introduce a heterarchical structure of decisions and, in particular, to describe general rules and special cases of behavior. In production systems, this enables creation of flexible decision systems [5], but such an extension of logic models may produce some undesirable effects. For example, introduction into a production system of a hierarchy and strategy of production usage that reflects the so frequent in human reasoning principle "exceptions to the rules are more important than the rules themselves" (e.g., in a novel by the French author Pierre Louÿs "Les aventures due roi Pausole" the everyday life of royal subjects was regulated by only two rules—"Do not harm anybody" and another, more general one, "As to the rest, do whatever you like") might cause the set of expressions generated by the production systems to be wider than that generated by algorithmic systems [6].

B. Of special importance in human reasoning are various scales onto which the reasonings are projected. In such reasonings as "We do not eat men—we are not cannibals" or "I am not like you, I always tell the truth", scales WE-THEY or I-YOU are used. These scales that are characteristic of the whole of human history, reflect the binary oppositions that underlie primitive thinking [7]: Our tribe and alien to us THEY; I belong to WE—to our tribe; WE are men, etc. But reasoning against similar scales is used even today.

Another use of scales in human reasoning is exemplified by Fig. 2 showing three scales: crimes, punishments and changes of correction as a result of punishment. Obviously, if punishment for theft is Q_1, i.e. $\alpha_1 \Rightarrow \beta_1$, that for murder should be greater, i.e. $\alpha_2 \Rightarrow \beta_2$, and $\beta_2 > Q_2$. Changes of correction, as may be seen from Fig. 2, are $\gamma_2 < \gamma_1$.

Reasoning on time or space also makes use of scales onto which events and facts involved in reasoning are projected [8,9]. Reasonings on frequency of events [10] or size of objects have the same feature. The fundamental role of scales in organizing the human "knowledge space" was determined by Osgood [11]. Studies that stemmed from his work have demonstrated that projections onto scales are one of the fundamental mechanisms of human thinking and reasoning. This has caused AI experts to introduce scales into logic models used for simulating human reasoning and decision making.

C. The scales in Fig. 2 commonly are ordinal, i.e. relations between events or facts over them are defined as fuzzy. It will be recalled that the main reason underlying the introduction of fuzzy sets by L. Zadeh was the need to allow for those features of human reasoning that are related to the use of fuzzy variables. They are exemplified by the following reasonings: "It often rains here, therefore, I take my umbrella almost without exception" or "Almost everybody who smokes cigars drinks brandy, but among brandy-drinkers there are many who do not smoke".

Logic models of reasoning with fuzzy variables use Zadeh's membership functions and special inference rules. Leaving aside the well known studies in this domain, we refer only to recent results that have extended syllogistics by introducing fuzzy variables [12,13]. Publications [8,9,10] apply to the scales fuzzy variables such as "soon, earlier, near, often, etc.".

D. In interesting studies of E. Berne and his followers [14,15] three characters were introduced—Parent, Adult and Child. Berne has demonstrated that the Parent's knowledge consists of cognitive structures that were learned by him without any criticism. For example, a man knows that one must not go to bed without taking off his shows, but never asks himself "Why?"; he was taught so from early childhood and remembers this fact forever. The Adult's knowledge has been obtained by himself, and prior to being learned, it was critically interpreted and checked. This knowledge relies upon confidence rather than belief. Finally, the Child's knowledge is impressions "without beginning and end" that are related to strong emotional experience.

Interestingly, the three characters that co-exist peacefully in any person, reason in different ways (this fact was missed by Berne). In the "Parent logic" such as "If A, then B", the premise A is somewhere in the remote past (Fig. 3). It has long been

Figure 1. Robot makes decision.

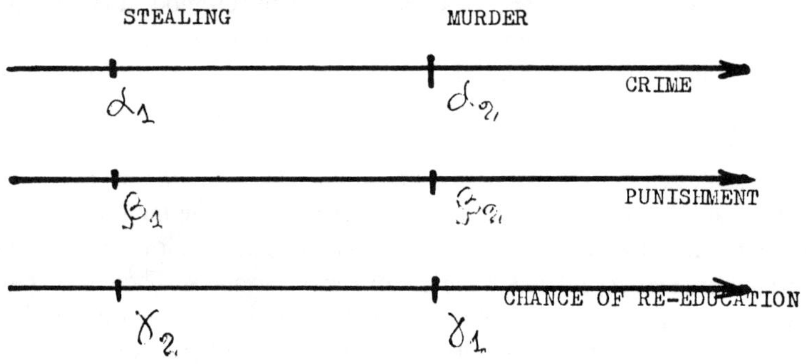

Figure 2. Scales in human reasoning.

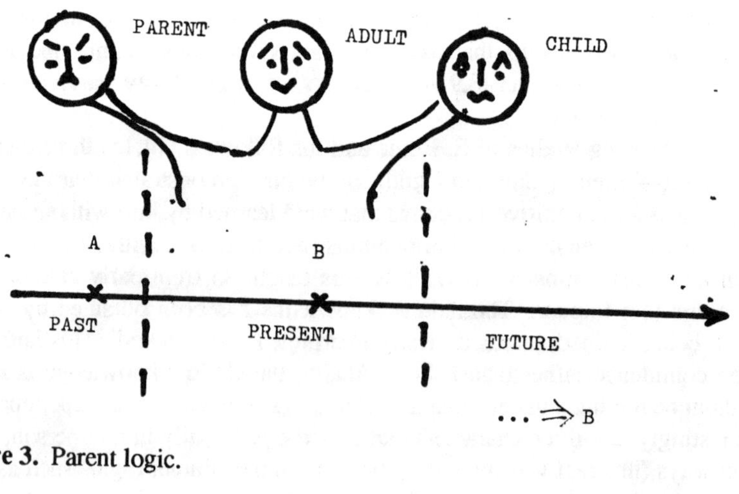

Figure 3. Parent logic.

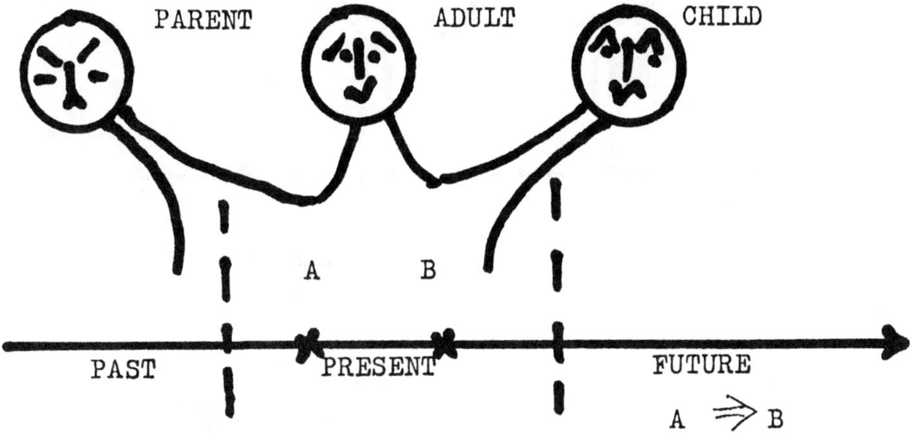

Figure 4 Adult logic.

forgotten, but is regarded as true because the conclusion is regarded as true. Reasoning of this type was especially characteristic of primitive society: "Before a hunt we dance in front of the fire in order to make the hunt lucky". The question "How is your dance related to the luck in the hunt?" received a standard answer, "I do not know. That is how our ancestors did it". In modern society "Parent logic" seems to be the one most frequently used. Many examples of "Parent reasonings" may be found in games and entertainments described in [14].

In an Adult's reasoning (Fig. 4) the cause A and consequence B are in the present. The truth of A is objective, and that of B may be readily verified now or in the near future. Reasoning of this type is well described by classical logic.

"Child logic" proceeds from the desire to have in the future some consequences B. It is concluded from "I want B very much" that the premise A is true (Fig. 5). If, for instance, a child wants a dog and the mother promises to buy it if the child will behave properly, the child may rightfully believe that he already behaves properly. A similar situation occurs when a person is very afraid of B and tries his hardest to convince himself and his associates that A is false. "Child reasoning" as well as "Parent reasoning" have a significant place in our thinking. One of the tasks of logic model designers for intellectual systems is to integrate these different reasonings into their models.

E. Classical logic pays attention mostly to models of deductive reasoning that are characteristic of strict scientific theories. In human reasoning, deductive models are rather rare. More frequently it features plausible reasonings [16] that are

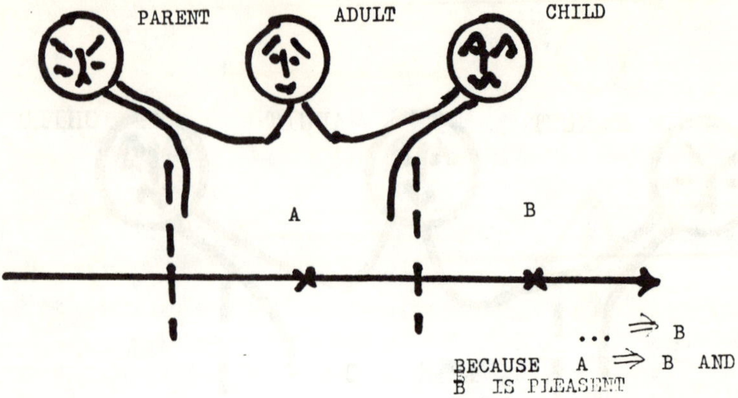

Figure 5. Child logic.

Figure 6. Two fate.

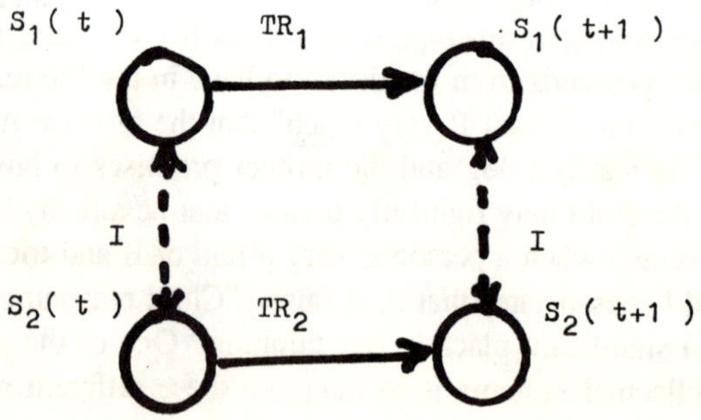

Figure 7. Transformation diagram.

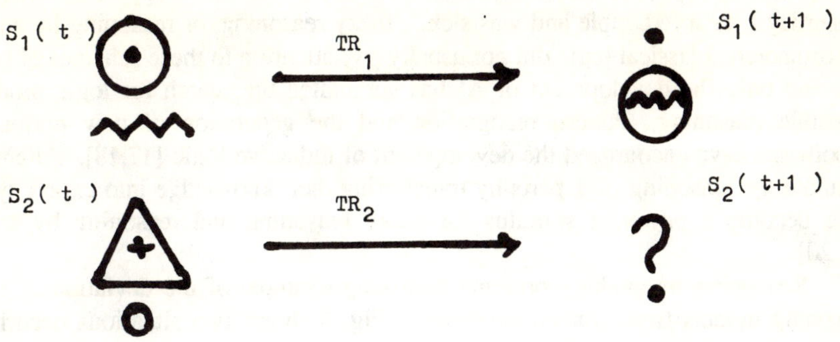

Figure 8. Example. Part 1.

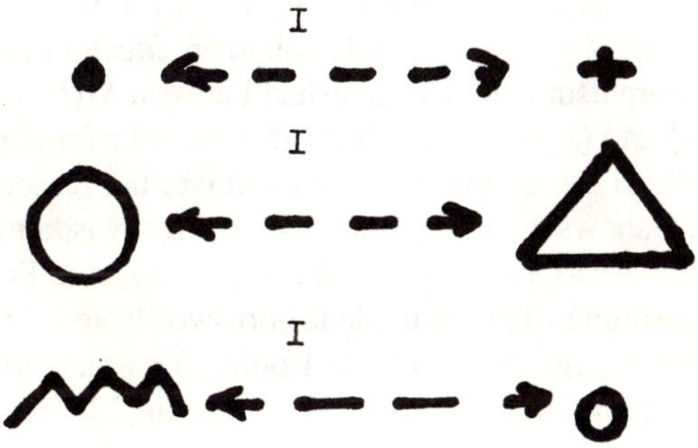

Figure 9. Example. Part 2.

Figure 10. Example. Part 3.

exemplified by inductive reasoning based on single events ("Red apples are dangerous. Yesterday I ate a red apple and was sick.", fuzzy reasoning, or reasoning by analogy. Unfortunately, classical logic did not usually pay attention to these schemes of reasoning, and only the development of AI has stimulated the search for logic models of plausible reasoning. Pattern recognition and the generation of new notions and hypotheses have encouraged the development of inductive logic [17,18]. Attempts to simulate the reasoning of experts by transferring their knowledge into expert systems have become a powerful stimulus for fuzzy reasoning and reasoning by analogy [19,20].

Reasoning by analogy presents a striking example of the deviation of logical reasoning models from conventional ones. Fig. 6 shows two situations occurring at some time t. There is some resemblance between them, but what? Is it possible to conclude something about the fate of Tristram and Yseult from their situation and that of Romeo and Juliet and the knowledge that at some time in the future Romeo and Juliet will die? The studies of reasoning by analogy rely upon an idea going back to Leibniz. According to this idea, a diagram as depicted in Fig. 7 is constructed where $S_1(t)$ and $S_2(t)$ are the situations shown in Fig. 6, $S_1(t + 1)$ is the situation where Romeo and Juliet die, and $S_2(t + 1)$ is a still unknown situation concerning Tristram and Yseult. If an isomorphism can be established between $S_1(t)$ and $S_2(t)$, it may be transferred to $S_1(t + 1)$ and ($S_2(t + 1)$ provided that the transformations TR_1 and TR_2 are isomorphic as well. If the diagram is commutative, the reasoning by analogy is regarded as true. Fig. 8 shows determination of X through reasoning by analogy. Isomorphism I between the descriptions $S_1(t)$ and $S_2(t)$ is shown in Fig. 9. Fig. 10 gives the final form of the reasoning. This example is borrowed from [21].

F. Human beings engage in reflexive reasoning for which no good logic model exists. Most advanced studies in this area are described in [22]. Attempts to construct such models have been made in [23] for collective behavior of "reasoning" automata.

Thus, the logic models required for AI systems differsignificantly from those covered by mathematical logic. The growth of new branches of knowledge such as the theory of types, nonmonotone logic, fuzzy logic and many others, testifies to its renovation and the desire to meet the requirements of AI systems.

REFERENCES

1. Kripke, S. A., Semantical Analysis of Modal Logic 1. Normal Modal Propositional Calculi, ZMLGM, vol. 9, pp. 67-96, 1963.
2. Prior, A. N., *Past, Present and Future*, pp. 1-168, Clarendon Press, Oxford, 1967.
3. Pospelov, D. A., Semantic Models in Psychology and Artificial Intelligence Systems, in *Human and Artificial Intelligence*, ed. F. Klix, pp. 45-58, North Holland Publishing Co., Amsterdam-NY-Oxford, 1979.
4. Pospelov, D. A., Situational Control, pp. 1-285, *Nauka*, Moscow, 1986 (in Russian).
5. Nilsson, N. J., *Principles of Artificial Intelligence*, pp. 1-372, Tioga Publishing Co., Palo Alto, California, 1980.
6. Kuznetsov, V. E., Mathematical Construction in K-systems, *Semiotica and Informatika*, vol. 27, pp. 62-81, 1986 (in Russian).

7. Frazer, G. G., *The Golden Bough*, pp. 1-829, London, 1923.
8. Litvintseva, L. V., Pospelov, D. A., Time in Robots and Dialog Systems, *Proc. of the Sixth IJCAI*, Tokyo, vol. 1, pp. 541, 1979.
9. Pospelov, G. S., Pospelov, D. A., Influence of Artificial Intelligence Methods on the Solution of Traditional Control Problems, in *Machine Intelligence*, vol. 9, ed. J. E. Hayes, D. Michie, L. I. Mikulich, pp. 331-348, Ellis Horwood Limited Publ., Chichester and John Wiley and Sons, N.Y.-Chichester, 1979.
10. Pospelov, D. A., Fuzzy Reasoning in Pseudo-physical Logic *Fuzzy Sets and Systems*, Vol. 22, No. 1/2, pp. 115-120, 1987.
11. Osgood, C. E., Suci, G. J., Tannenbaum P., *The Measurement of Meaning*, pp. 1-275, Urbana, 1957.
12. Chesnokov, S. V., The Effect of Semantic Freedom in the Logic of Natural Language, *Fuzzy Sets and Systems*, Vol. 22, No. 1/2, pp. 121-154, 1987.
13. Efimov, E. I., Determinational Calculi, Izvestija AN SSSR. *Technicheskaja Kybernetika*, No. 5, pp. 62-81, 1986 (in Russian).
14. Berne, E., *Games People Play*, pp. 1-192, Grove Press, Inc., NY, 1967.
15. Harris, T. A., *I'm OK—You're OK*, pp. 1-269, Pan Books, London and Sydney, 1973.
16. Polya, G., *Mathematics and Plausible Reasoning*, pp. 1-527, Princeton Univ. Press, Princeton—New Jersey, 1954.
17. Plotkin, G. D., A Further Note on Inductive Generalization in *Machine Intelligence*, vol. 6, ed. D. Michi, pp. 101-124, Edinburgh Univ. Press, Edinburgh, 1971.
18. Hajek, P., Havranek, T., GUHA-80—An Application of Artificial Intelligence to Data Analysis, *Computers and Artificial Intelligence*, No. 2, p. 107-134, 1982.
19. Chouraqui, E., Construction of a Model for Reasoning by Analogy, *Proc. ECAI-82, Orsay, France, pp. 48-53, 1982*.
20. Farreny, H., Prade, H., About Flexible Matching and its Use in Analogical Reasoning, *Proc. ECAI-82*, Orsay, France, pp. 43-47, 1982.
21. Pötschke, D., Toward a Mathematical Theory of Analogical Reasoning, *Proc. ECAI-82*, Orsay, France, pp. 54-59, 1982.
22. Lefebvre, V. A., *Algebra of Conscience*, pp. 1-194, D. Reidel Publ. Comp., Dordrecht-Boston-London, 1982.
23. Gaase-Rapoport, M. G., Zakharov, V. N., Pospelov D. A., The Influence of Nonuniformity in the Models of Automata Collective Behavior, *Proc. IJCAI-75*, Tbilisi, USSR, vol. 1, pp. 321-327.

DISCUSSION OF POSPELOV'S PRESENTATION

Mitroff

We *simultaneously* reason across parent, adult, and child "logics." Do you agree?

Pospelov

I agree that all three types of logic are accessible to each individual simultaneously, but at any given moment he can use only one. In multistep reasoning or behavior we, as a rule, have a mixture of all three types of logic.

Mitroff

Have you worked with logics other than Berne's parent, adult, child model?

Pospelov

Yes. This was just one study.

There are many kinds of rational and irrational logics. There are parent and child irrational logics. There is also a difference between psychological time and space on the one hand and physical time and space on the other hand. We move events into the past in order to forget them and bring them into the present in order to anticipate or make forecasts.

Apokin

Don't you think that multivalued logics would be more appropriate for AI?

Pospelov

Even endless valued logics.

Apokin

Fuzzy algorithms, for example?

Pospelov

Yes.

Lavrov

I think that we must develop "women's logic" together with your "parent logic" and "child logic."

Pospelov

Actually "women's logic" is not manifested in just one step of a set of the rules or in a conclusion of the type A leads to B but rather in shifts in rules. For example:

1) He came back late.	He was somewhere.
2) He was somewhere.	He was with another woman.
3) The other woman.	She is better than me.
4) She is better than me.	This cannot be.
5) He came back late.	That is bad, but he still loves me.

Here rules 1 and 3 are adult logic, rules 4 and 5 are a child's logic, rule 2 is parent logic, and all together constitute "women's logic."

Von Glasersfeld

You speak of different types of logic and attribute one of them to children, another to parents. I have the impression that these two types of reasoning often occur in one and the same person. Do you have any hypothesis as to what would make a person switch from one logic to another?

Pospelov

I think that all people use all three types of logic. In the process of communicating with one another, people rationally or intuitively switch from one type of logic to another depending on their aims, psychological state, and resources (for example, of time).

Sadovsky

Are there some rules for switching among the types of logic? If so, then this activity is rational as well as traditional logical behavior. Our concept of "rationality" seems to be evolving.

Is it possible to apply the term "logic" to nonrational activity? I believe that "nonrational logic" is a contradiction in terms.

Pospelov

I use the term "logic" in its broad sense—not only as a set of rules of reasoning but also as rules for behavior. If the rules of switching from one type of logic to another (e.g. from parent to child logic) were verbalized, then behavior would be rational as would be the logic of this behavior. In the opposite case (and this is how, I think, it is) the logic of behavior will be irrational.

Klir

What are the current research activities in the area of fuzzy set theory in the Soviet Union?

Pospelov

In the USSR we have long been interested in the work of Lotfi Zadeh. What was attractive was the principal goal of work in the areas of fuzzy sets and fuzzy reasoning, i.e., to create a formal apparatus making it possible to describe the human qualities of reasoning and behavior. Unfortunately the work in this area very rapidly took on a conventional mathematical character. Fuzzy algebra, fuzzy topology, etc. started to emerge. The basic goal was forgotten. In the USSR many of the specialists in large systems management are researching fuzzy set theory in the purely mathematical sense. However, some management specialists and AI specialists are trying to get back to the original goal of fuzzy set theory. In particular, Soviet specialists are developing special pseudophysical logics (temporal, spatial, etc.) using fuzzy scales and fuzzy inference rules making it possible to intuit human reasoning.

Ben-Eli

Have you made use of Gordon Pask's work on conversation theory?

Von Glasersfeld

Pask's "conversation theory" can work only if the interaction goes in both directions. My question to Pospelov was triggered by the idea that politicians, who usually purport to play a parental role, quite often employ "child logic" in their own thinking.

Lapin

Usually by "rational" behavior we mean "purposeful" behavior. But recent research has shown that "purposeful" behavior is only one form of rational behavior, while value-guided/normative and traditional behaviors are also rational. What is your understanding of rational behavior?

Pospelov

I understand the term "rational behavior" somewhat differently. If behavior can be verbalized (i.e., I can explain it to myself or to others in words), then for me it is rational. Therefore, consciously value-guided/normative and traditional forms of behavior are rational. What in behavior cannot be verbalized is not rational.

CHAPTER

9

SPONTANEOUS SOCIAL ORDERS†

Milan Zeleny

As the results of man-engineered experiments with social design, social "revolution", socialist "perestroikas", and other feats of "social engineering", are crumbling down, they are causing large-scale human suffering through their failures. There is a renewed awareness that self-organizing and spontaneous properties of complex social systems are much too powerful (an much too vulnerable at the same time) to respond or be exposed to the endless, reductionistic "tinkering" of policy "makers", "scientists" of the artificial, and "engineers of human souls".‡ The mankind is again ready to learn how to "trigger", "catalyze", "sustain" and "lead-manage" a *spontaneous process* of social self-organization, it is becoming less inclined to design another "central super-controller", "information-processing command system", or "World Brain".

---

†This paper was first presented at the United Nations University Symposium on the *Sciences and Praxis of Complexity*, in Montpellier, France, May 9-11, 1984. The author would like to acknowledge the U.N. University's generous support as well as the permission to publish. Reprinted also by permission from *Int. J. General Systems*, 1985, Gordon and Breach Science Publishers, Inc.

‡Even Geoffrey Vickers, in his *Human Systems are Different*[26] ends up using a similar metaphor: "It [his book] does not offer a blueprint for the design of human societies. I hope it will weaken the contemporary urge to regard such activities as akin to engineering rather than (at most) to gardening." One can only change the "at most" to "at least".

The purpose of this paper is to show: (1) that theories dealing with "spontaneous social orders" have deep historical roots and (2) that systems sciences are in a good position (better than economics, engineering or sociology) to build upon these roots and expand the theories into useful, practical methodologies. For example, modern theories of *autopoiesis and order through fluctuations*, especially their rich, computer-based simulation experiments, provide a good and solid foundation.

1 INTRODUCTION

We live in the age of large-scale social and organizational *designs*, artificial intelligence, social engineering, and even "sciences of the artificial". The ordering activity of an ordering mind seems to be lurking behind it all—or so it may appear to a less inquiring mind.§

There are, of course, man-made, artificially created societal orders and arrangements. They are achieved by arranging the relations between societal parts *according to a preconceived plan*: organization charts, management hierarchies, command economies, artificial cities, military systems, political "machines", orchestrated movements, specifically designed institutions, concentration camps. The power of many men can be increased by such deliberate coordination of their efforts and many of the crowning achievements of man rest on the use of this technique. It is an order we all understand because we know how it was made, how it came into being.

> "But it is not the only nor even the chief kind of order on which the working of society rests; nor can the whole of the order of society be produced in this manner.
>
> The *discovery* that there exist in society orders of another kind, which have not been designed by men but have resulted from the action of individuals without their intending to create such an order, is the achievement of social theory."

The above observation comes from F. A. Hayek.[10] Needless to say that this "discovery", dating from Adam Smith's "invisible hand", has not yet led to serious study and competent insights into the underlying self-organizing mechanisms producing spontaneous social orders. The "Invisible Hand", if anything, has remained invisible.

K. E. Boulding[8] characterizes the invisible hand as the concept of the development of social order through the *unplanned* interactions of large numbers of individuals, each seeking their own gain through production and exchange, coordinated through the development of the system of relative prices. "This is a considerable conceptual advance beyond the rather one-dimensional mechanics of Newton," adds K. E. Boulding. The whole of modern economics and its systems of equations is based on the Newtonian mechanics.

---

§In a remarkable report,[9] Albert Hajnal puts a finger on the curse of modern science: "What we have to deplore is not so much the fact that *scientists are specializing*, but rather the fact that the *specialists are generalizing*."

Social systems (of the human kind) are a mixture of deliberate arrangements of man-planner-designer, interacting with complex, spontaneously emerging orders, increasingly complex and important, made by no one, preconceived by no one, foreseen by no one. Both aspects—artificial and natural, man-made and spontaneous, designed and self-produced, simple and complex—should be studied in their interaction and mutual determination. The following analogy is useful: one can grow a big, healthy tree or one can take some wire and scissors and impose a formal structure upon it to grow a bonsai.

2 HISTORICAL SYNOPSIS OF SPONTANEOUS SOCIAL ORDERS

The exquisite observations of *Adam Smith* in his *Wealth of Nations* have already been mentioned. He perceived, in the seeming chaos of the marketplace, always-emerging spontaneous order, an order generated through self-correcting actions of the participants. Smith realized that seemingly inchoate flux of phenomena and events possesses self-ordering properties well beyond the wildest dreams of most assertive social engineers; he first realized that *chaos is unstable* and more; that chaos is a particular structural manifestation of an order not yet perceived or understood by human observers.

Unraveling the underlying self-organizing mechanism should be the rock upon which economics is built.

Instead, according to Kristol:[16]

"The economic perspective can now provide a series of static "snapshots' of a universe-in-change, not a discursive description of a dynamic marketplace in which men learn, innovate, and generate economic growth. Economic *states* become the focus of analysis as contrasted with the economic *process* that Smith rather casually (through brilliantly) analyzed."

So, even today, we remain as casual as Smith was about the concept of utmost importance; we study equilibrium instead of disequilibrium, statics instead of dynamics, structural manifestations instead of self-organizing principles, perfect instead of imperfect competition, being instead of becoming, mechanism instead of organism, artificial and man-made instead of spontaneous social orders, simple instead of the complex.

We might be replacing the Newtonian clockwork model by a cybernetic model, mechanical statics by mechanical dynamics, or systems of linear equations by systems of differential equations—but the radical *conceptual* revolution which would allow us to see social systems as self-organizing and self-producing systems is still not in sight.

Adam Smith was not of course alone in his perceptions. Giambattista Vico[4] and Bronislaw Trentowski[25] (in his *Cybernetyka*, 1843) have understood societies as organisms. Interested reader should also consult Refs 27 and 28 for a review. We shall start with Carl Menger.

Carl Menger[19] was the first scientist to use the term "spontaneous organization" in the sense explored here. He noticed that the admirable functionality of all parts of natural organisms, with respect to the whole, is *not* the result of human calculation. The same seems to hold true for numerous social institutions: they present themselves to us as "natural", i.e. not artificial, products, and as *unintended results of historical development.* One needs only to think of such complex phenomena as money, law, language, the origin of markets, communities, states, etc.

Menger published his major statements in 1883, just naturally preceding Jan Smuts and his "holism" which expressed similar philosophy. The analogy of social phenomena and natural organisms, as drawn by Menger, refers only to those phenomena which are *unintended* product of historical development; the rest are the result of human calculation and thus are not comparable to *organisms*, but to *mechanisms.* Social institutions are therefore mixtures: neither artificial nor natural, both intentionally created and spontaneously generated.

Thus, according to Menger, social systems are best described as *social organisms*, i.e. not natural organisms and not social mechanisms. He poses the following question: "how can it be that institutions which serve the common welfare and are extremely significant for its development come into being without a common will directed toward establishing them?" He cites social prices of goods, interest rates, rents, wages, etc., as examples of such institutions. His analysis of the origin of money and of the origin of law represent excellent insights into spontaneous social orders. Similarly, development of new localities, separation of professions, division of labor and even origin of the state are analyzed in detail.

Although Menger laid direct and rich foundations to the work of F. A. Hayek, two great thinkers contributed to the concept in the meantime: Jan Smuts and Alexander Bogdanov.

A. A. Bogdanov[6] published his Tektology in 1912. In this general science of organization he formulated basic organizational rules that govern the origin (creation), maintenance, and destruction of systems. Basic rules of conjunction (production) and disjunction (disintegration) are interconnecting components into network and, through their interaction a tektological boundary *emerges.* (Observe that boundary of the systems and its structure are not given but come into being, emerge).

Bogdanov was the first to recognize the difference between organizational rules (guiding the interaction of components) and their structural manifestation (spatio-temporal pattern, structure). This distinction later reappeared and brought its fruit in the theory of autopoiesis.[29] Until today, however, much more attention is being paid to pre-existing, fixed structures and too little to the underlying organizational rules which bring such structures into existence.

Jan Christiaan Smuts is known as the father of holism[21] which he introduced in 1926. Smuts states:

> "An organism is a continuous *autogenesis.* It can only be explained by reference to its past and its future as well as its present: the central *structure* is not sufficient and literally has not enough in it to go round in the way of explanation.

Whole is always transcendent of its parts and its character cannot be inferred from the character of its parts."

By introducing holism, Smuts draws attention, albeit only of a few, to the fact that in organisms (as opposed to mechanisms), both natural and social, not only the whole cannot be inferred from its parts but the parts themselves can only be known within the ever-changing whole: the whole and the parts mutually determine one another and are related in a circular dependency.

F. A. Hayek,[11] building upon such excellent foundations, introduced the most extensive argument for the study of spontaneous social orders. He recognizes "the rules of conduct of the individuals" as leading to "the production of the overall order" which is not "the conscious aim of individual action". In the modern vocabulary of autopoiesis, one would say: organization of components and processes of their production gives rise to their structure (their particular spatio-temporal arrangement) in a natural, self-organizing fashion.

Hayek thus is aware that the same overall order of actions may be produced by different sets of rules of individual conduct (different organizations can manifest themselves in the same structure). He also recognizes the role of environmental embedding: the same set of rules may lead to different orders of action under different external circumstances. He sees clearly the holistic closure: a certain kind of overall order may contribute to the survival of the members of the group. He also understands Bogdanov's role of evolutionary selection: the evolutionary selection of different rules of individual conduct operates through the viability of the order it will produce.

Hayek thus offers a crucial, modern insight into behavior of social systems: *the rules of conduct will always act only as a constraint on actions induced by other causes.*

Thus, he does not deny the role of human will, purpose and choice; he recognizes the needs, objectives and goals as stimuli and impulses for action—to be constrained and streamlined by the rules of conduct (in biological systems a rule of conduct itself may be a sufficient cause of action).

It is to be concluded that the most potent precursor of modern theory of autopoiesis has its roots in theories of *social systems*. It is in studying social systems where the spontaneous self-production was first observed and studied. In fact, all biological systems are—and should be studied as—spontaneous social systems.

It is interesting to note that social orders can persist in spite of the continual, even random, replacement of components: imagine a dictator who specifies all particular actions to be performed but keeps selecting the individuals to perform them by random drawings—there will exist and persist a recognizable order.

Examples of spontaneous social orders would include the arrow formation of migrating wild geese, the defensive ring of the buffaloes, or the hunting formation of a pride of lions: it can be safely presumed that there is not an awareness of the whole on part of the individual. There are rules of responding to the immediate environment which constraint and coordinate actions of independently acting individuals. More complex examples are insect societies of bees, ants, and termites, characterized by complex structures, "sophisticated" division of labor, and appropriate evolutionary adaptations. (To study such behavior through genetic transfer of encoded informational "blueprints" is thus misplaced and will cause continuing ignorance about the innate rules of conduct of the individuals).†

Thus, human organization (rules of conduct) is evolving in response to the changes in the environment (circumstances). An environmental perturbation requires, if the whole is to persist, a change in the order of the group (structure) and, consequently, a change in the rules of conduct of the individuals (organization). The sets of rules of conduct will evolve only as wholes because the selection process of evolution can operate on a given order only as a whole, not on its parts. (We observe the same in biological systems although our theories of evolution are based on different, atomistic notions).

Hayek's theory of social systems, especially as expressed in Ref. 11, is complete and thoroughly modern. He insists that economic and linguistic theories are endeavoring to reconstruct from regularities of the individual actions the character of the resulting order. Yet, the nature of the different kinds of rules of individual conduct is frequently left obscure. Then we cannot study the question of which of these rules can be deliberately and profitably altered, and which are better left to gradual, natural evolution—we cannot evolve a meaningful *theory of action*.

Societies differ from individual (complex) systems by the fact that their components are themselves complex systems whose change to persist is enhanced by (or dependent upon) their being part of this more comprehensive system (society). *All living systems form societies*. Individuals, in their conduct, are constrained by the very environmental structure (society) they form. They accept such constraints in order to preserve or enhance their individual survival. By doing that, they further strengthen the structure (society). No "design by a maker" needs to be evoked.

Individuals of a group of hungry animals act individually and independently to satisfy their own hunger. Through doing that, because they possess common set of responses to a given environment (as a result of identical physiological structure and similar experience), they generate a precursory formation which becomes an environmental constraint to them. They adjust their actions to this emerging social order (if they do not, they not only destroy the order but perish as individuals), ultimately forming a complex order of apparent cooperation, communication and collective social action. This is Adam Smith's "invisible hand".

From the above thoughts emerges Ortega y Gasset's observation, "I am myself and my circumstance", as a clear and deep insight into the human condition.

Humans, in their history, have shown distinct preference for observing the rules of conduct whose consequences are predictable rather than those of unpredictable consequences. This, in itself, provides an engine for evolution of social orders.

†So called "biological analogy" is quite powerless if biological organism itself is viewed only as a machine (a computer). See for example discussions in Refs 14 and 23. Neither biological nor social systems are machines and should not be studied as machines (or "information"-processing computers).

Predictability of consequences is much more powerful criterion for choosing an action than their relative "utility" or "morality". A criminal who is persistently and without guilt violating man-designed and authority-imposed rules of conduct might never even consider raising his voice to his father, striking a child, or betraying his friends. The world becomes frighteningly unpredictable if one deviates from the spontaneously evolved, natural rules of conduct. To know what is taboo is as important as any positive knowledge about the environment. "Thou shalt not", by conveying the *sense of constraint*, is as important rule of conduct as any positive guide to action.

We should not "plan" a society *because* it has grown too complex; in fact, the order of such complexity can only be preserved if we aim at the formation of a spontaneous order. We cannot improve upon a spontaneous order by issuing direct and specific commands to its individuals—no central authority can ever know "*their* circumstance".

The interpretation of law as an instrument of arrangement (konkretes Ordnungsdenken) is far and distinct from its interpretation as a constraint on the conditions of spontaneous order formation, and thus far from Hayek's interpretation of social systems. A complex society can be effectively controlled through the conditions of the formation, not through direct design and deliberate arrangements. That approach is suitable only for relatively simple orders of limited creative capabilities (although very efficient in destruction: concentration camps, military complexes and police bureaucracies).†

The counterintuitive dilemma for some is that in an apparent striving for rationality in the sense of fuller taking into account all the foreseeable consequences, we may achieve even greater irrationality, less effective taking into account of remote effects. In the words of von Hayek:[11]

> "...the consequence of modern democratic legislation, which disdains submitting to general rules and attempts to solve each problem as it comes on its specific merits, is probably the most irrational and disorderly arrangement of affairs ever produced by the deliberate decisions of men."

Thus we "solve" the problem of hunger by shipments of food, the problem of racial integration by forced busing of children, the problem of homeless by providing

---

†The "engineering" view of control, as emphasized by Ashby's "Law of Requisite Variety", does not apply to social systems. "Only variety can destroy variety" implies that the "control" must be able to respond differentially to all the states the controlled unity may display. Thus, the controller must be at least of the same complexity as the controlled. Ashby himself was quite aware of this "trite law" being a fallacy with respect to social systems: "...the law says that his [Hitler's] control over Germany amounted to just 1 man-power, and no more. (Whether this statement is true must be tested by the future,...)" [W. Ross Ashby, *An Introduction to Cybernetics*, John Wiley & Sons, New York, 1966, p. 213].

The future has tested that, many times. In social systems, the extent of control has something to do with the willingness, agreement or consent of the controlled to be regulated. This is addressed by the concept of "leadership" which is missing from the cybernetic theory of control.

It is no longer possible and more, unethical, to understand society as a kind of machinery in need of regulation.

shelters (not homes), the problem of poverty by providing handouts, the problem of "Poland" by billions of dollars, and so on. We never concern ourselves, hopefully because of ignorance and not politics, about the causes, the spontaneous, ever-renewing and self-maintaining conditions, of human misery.

Societies based on democratic (governmental) legislation, *not* democracies, have taught their citizens to shift their natural responsibilities, as members of human communities and societies, on the shoulders of central governments. The larger the shift, the higher the price (the taxation level). It is not difficult to outlaw poverty, to ban unemployment, to forbid crime—it is only costly and it effects the structure, not the underlying organization which generates the phenomenon. Thus, this approach is doomed to failure.

3 PARALLEL VIEWS ON SOCIAL AUTOPOIESIS OR SELF-PRODUCTION

The intellectual history of spontaneous social orders has, of course, been interrupted and it virtually ends with von Hayek.† The idea itself has sometimes been independently generated and rediscovered, in different fields and in different domains of human inquiry.

Alain Touraine[24] published his "The Self-Production of Society" in 1973. Although working without the benefit of Hayek's powerful insights, Touraine does understand society as a network of actions and relations and as being totally the product of *social relations*. In this sense,

> "Society is not just reproduction and adaptation; it is also creation, self-production."

Self-production (or autopoiesis) implies that a society, first and foremost, is an agent (and object) of its own self-production. Touraine also employs the rules of conduct of individuals as a generative force towards unintended, spontaneous social arrangements:

> "The meaning of conduct must be explained not by the consciousness of the actor or by the situation he is placed in, but by the social interactions in which he is involved."

The concept of closure and self-reference is also not foreign to him: societies act upon their own functioning, they turn back upon themselves. Touraine also goes beyond postulated invariance of the underlying rules of conduct (organization) and, as Hayek, fully understands that these also evolve, and that the system involved is not only self-producing but also self-organizing:[24]

---

†Von Hayek was even able to formulate the fundamental problem of economics in the remarkably modern framework of *knowledge*. "How can the combination of fragments of knowledge existing in different minds bring about results, which, if they were to be brought about deliberately, would require a knowledge on the part of the directing mind which no single person can possess?" (F. A. von Hayek, "Economics and Knowledge", *Economica*, February 1937, p. 52).

"Human society cannot be reduced to an organism always reproducing itself immutably, dependent for any transformation in itself upon mutations brought about by internal or external events. Its capability is not limited to adapting itself to a changing environment and is constantly modifying the rules by which it functions."

This is another insight from social sciences which should find its manifestation in our interpertation of biological systems. The organization of a living system, far from being simply "invariant", continually evolves and adapts itself (self-organizes itself) in response to environmental perturbations and thus induced structural changes.

Alain Touraine says it simply and tersely:

"Society produces itself."

Nobody else ever put it so boldly and with such confidence.

Touraine goes, of course, further: society not only produces itself, but it also adapts itself and consumes itself. It concatenates the processes of production, persistence and disintegration (consumption) circularly upon itself. Touraine identifies the corresponding areas of inquiry as "the field of historicity", "the institutional system", and the "social organization".

Touraine inches toward interpretation of the exhaustion of industrial society not as being replaced by a society of equilibrium but by a society in which society's capacity for action upon itself as a whole is making a qualitative step.

Industrial society does merge into the stasis and equilibrium of "postindustrial society"—as a simple projection of existing trends—but transforms its own organization (the underlying rules of conduct) in response to limitations and constraints it itself has created. The society acquires not only new structure but, more importantly, new organization (new rules of conduct). This is evident in emerging decentralization of work, increasing work at home and self-service, breakdown of hierarchies in high-technology businesses, and the reversal of the process of division of labor. (All this is different from a "structural" argument that more people are involved in services and information industries—thus "post-industrial" society).[31]

Another independent, parallel view comes from one of the proponents of autopoiesis theory, Humberto R. Maturana.[18] Although a biologist, Maturana has expressed unambiguous and competent views on social systems and human condition. He states, as confidently as Touraine:

"A social system is a biological system."

The label "biological" must be understood in Maturana's sense: autopoietic, self-producing, living. That is *not* as a genetically reproducing machines.

The only other statement of similar unambiguity comes from a management and organizational scientist, Stafford Beer:[3]

"Yes, human societies *are* biological systems."

It should be obvious that if any of the assertions presented so far are even partially correct, then studying social systems as machines or mechanisms (or *artificial* contrivances) is bound to be misplaced. There is bound to be natural resistance to the "biological" view of society; even from biologists themselves.

The issue is philosophically complex and we shall not attempt to resolve it here. Let us simply state that whether or not something is or is not labeled as "biological" is in itself irrelevant. It all depends how we understand "biological" at a given stage of history; as a complicated machine, chemical or electric factory, sophisticated super-computer, or self-organizing autopoietic system? Let us simply say that living systems cannot be understood and should not be studied apart from social systems they create and on which they are dependent as individuals.† There are no individuals apart from social systems and no social systems apart from individuals; these are two aspects of the same "living" system.

Let us return to Maturana. He expresses the same idea by observing that a social system can only be integrated by living systems, and the relations and interactions between living systems that do not involve their autopoiesis do not participate in its constitution. In other words, Maturana sees social system as a network of interactions and relations, a medium in which human beings realize their autopoiesis. He introduces the notion of a *natural* social system. This is important because humans are capable of creating *artificial* "social" systems, especially in this age of "the sciences of the artificial". Concentration camp is neither autopoietic nor spontaneous, it is not self-maintaining and does not emerge from the underlying rules of conduct. Yet, it is a "social" system. It has been created by man as a designer and planner, externally, through operating directly on the structure. Although such artifically imposed structure affects the rules of conduct of the individuals, the rules of conduct themselves do not generate the structure. Remove the pressure of the "designer" and the whole social system disintegrates: everybody goes home.‡

In some other statements of Maturana we hear echoes of Bogdanov and his categories of positive and negative selection:

> "Social system is a medium which operates as a recursive *positive selector* of those properties of its components that confirm it (social system), and as a *negative selector* of those which negate it.

On a purely ethical basis and in response to some criticism of his views, Maturana argues that "no human being can be effectively replaced" in a given system—simply because no two human beings are identical. This is a weak argument because what is striking and important about natural social systems is their ability to maintain identity, continuity and autonomy in spite of (or because of) even a substantial turnover of human individuals. A family, village, town, market, etc., may retain their characteristics and definitional features over centuries and countless generations.

†Biological and social systems, for example, do not have goals or objectives, only individual humans do. Here the fallacy of "biological analogy" becomes clear; because organizational goals cannot be satisfactorily defined in principle, some writers[15] conclude that "...if this is so, organizations do not look much like organisms". Observe that this assumes that organisms do have goals.

It is much more useful to understand that so called "goals" of biological organisms are anthropomorphic imputations (assignations of the human observer). As Keeley[15] points out, we have to distinguish goals *of* an organization from goals *for* an organization.

‡People scale Berlin walls, flee in raincoat-made balloons, forge documents, and plain simply risk their lives in order to escape one or another carefully "engineered" and well designed social "paradise". This also implies that humans can compare, quite reliably, different cultures and ways of life.

The ethical issue, if any, concerns the wisdom of perpetuating the myth of man as a creator, designer and master of human society—when he is plainly not. Of course that individuals are and always have been replaceable in the great schemes of self-renewing social systems. It is precisely this knowledge which makes us effectively human and *useful* to other humans within our social system. Most miseries and disasters of human race stem from the pathological condition when certain individuals or groups viewed themselves as "irreplaceable".

Man as a planner, man as a designer has redesigned rural economies, ignored peasant needs and customs, clobbered economies with foreign aid misspent on industry and mechanization, imported Scotch whiskey and German Mercedeses instead of knowledge, fought rebels with starvation, forbade Russians to drink, designed import taxes on humanitarian aid, allocated 100 million on military parades, built convention centers and put flags all over the countryside. When millions of children started dying, the designer mobilized emergency airlifts bringing heaps of food and blamed the problems on—the weather. Man, as a planner, never reflected on the organizing principles of a society, stimulus and motivation for self-reliance, self-sufficiency, people and their relationships, thus ignoring or damaging spontaneous self-organizing forces of human societies.

Touraine and Maturana are quite solitary in their views on self-production of society. That should not be interpreted as that others do not, at least in part, find the "spontaneous" or "autopoietic" way of thinking useful. Even Hagerstrand[12] finds it useful to study streams of human beings "flowing through" a fairly rigid organizational structure. Boulding[8a] reminds us that most efficiently man-designed structures are military systems (destruction and disintegration rather than creation and production). Man keeps fighting spontaneous orders (free markets, underground economy, money, etc.) but very little the man-designed structures and institutions.† Luhmann[17] considers *events* to be basic units of social systems and their dissolution and disintegration as necessary for reproduction of social systems. Atlan[2] distinguishes between *artificial systems* (finality imposed by the planner) and *natural systems* (finality unknown, generated by the system itself). Holling[13] recognizes fundamental continuing disequilibrium of social (ecological) systems: new substrate → production → aging → degradation → new substrate. Thus, complex social systems exhibit enormous *resilience* to perturbations and to *management*.

Allen[1] shows very convincingly, in his computer simulation studies, that each social actor, taken separately, may have very simple and different decision criteria and desires, but the dynamic "folding" of the system can give rise to unintended and complex patterns and flows.

Prigogine[20] refers to a widespread interest in, and recognition of, the vital importance of the creative *self-evolution* of complex socio-technical systems. The global features of such systems do not reflect the fulfillment of a "master plan", but rather an evolution of process and structure in directions not necessarily planned, intended or

†Recent successful dissolution of regulatory CAB (Civil Aeronautics Board) in the U.S.A. remains an exception, yet the benefits to the quality of life are immediately felt.

desired by any single actor or group of actors. He even speaks of the emergence of "eco-science" as an essential element in the new dialog of man with nature.

4 THEORY AND MODELS OF SPONTANEOUS SOCIAL SYSTEMS

Social world is in a continuous *state of becoming*, it is therefore incomplete to model it as if it were in a *state of being*. Maturana's conception of social system (human or animal) is useful because it allows us to distinguish between organization and structure of a social system:

Social organization is a network of interactions, relations and processes through which the living components maintain their autopoiesis and thus produce, maintain and renew the same network of interactions.

Minimum processes involve those of production, linkage, and disintegration. *Production* refers to the rules and relations guiding the entry (birth, membership, acceptance, etc.) of new living components; *linkage* guides the association, roles, functions and positions of individuals during their tenure within a given organization; finally, *disintegration* refers to the processes associated with termination of membership (death, separation, expulsion, etc.).

All three types of processes must be present and balanced if a given organization is to exhibit self-production, self-maintenance and self-renewal.

Social structure refers to a particular contingent of living components integrating and constituting the organization at a given time and place. Structure is a particular spatio-temporal pattern of particular individuals: a "snapshot" of a state of being of the underlying world in a state of becoming.

Social system consists of the generative social organization and its particular manifestations and embeddings: social structures.†

Example. Family and *club* are examples of two different social systems. Their organization is different: they differ in the rules of conduct guiding the entry, roles and positions, and termination of their members. Individuals, through their conduct and adherence to (or disobedience of) rules, realize their own autopoiesis and thus recursively produce the same organization that produced them (as members). In a family, for example, a child is born not only as a biological entity, but also as a social entity: son, heir, citizen, etc. A newborn automatically enters a number of social systems and takes part in their autopoiesis. A club functions on different rules of conduct but its structural arrangement may or may not resemble that of a family. In the minimum, it could have the same individuals in common with a given family. The roles, hierarchy of power and the form of communication could resemble those of a family.

---

†Hajnal[9] for the first time makes a distinction between organization and structure as applied to the concept of *culture*. Instead of simply listing the attributes of culture as a static construct, he concentrates on spontaneously formed culture-*generators*: (1) The reproduction of society, (2) Innovation, (3) Productive work, (4) Organization of work, (5) Exchange and interest relations, (6) Organizations: communities, associations, federations. These *generators* are remarkably invariant (like organization in autopoiesis), they are not artificial products of man, but originate and renew themselves spontaneously. *They* are the fundamentals of culture.

Given the above, we can distinguish two basic kinds of *social change:* organizational and structural.

Organizational social change refers to the change (whether spontaneous or imposed) in the underlying rules of conduct. If, under such change, the individuals remain capable of maintaining their autopoiesis, they will recreate the changed organization together with its new structural manifestations—the change will be viable, long-lasting, capable of its own self-renewal. Otherwise, the change may lead to the demise of a given social system as a self-maintaining entity.

Structural social change refers to the change (whether spontaneous or imposed) in the outward structural manifestations (changes in personal contingent, reassignment of roles, restructuring of hierarchical linkages, and spatio-temporal displacement). The underlying social organization is unaffected. This type of social change, if not brought about by the underlying organizational process, is not self-maintaining, is short-term, and must be externally enforced and policed. The underlying social organization will continue producing its own structures (necessary for its own autopoiesis and the only ones it is capable of producing). Depending on the severity of structural change (e.g., physical or social destruction of most individuals) or the intensity and longevity of its enforcement (e.g., policing geographical location), the underlying social organization will either cease to exist as an autopoietic unity or will manifest its autopoiesis in a pathological (either predatory or deprived) pattern.

Structural change, if not severe, is not capable of changing social organization and thus is insufficient as a tool for social change. To be effective in a long run and in terms of social self-production, the underlying social organization must be changed first.

Examples. (1) Free price formation continues in centrally controlled economies in spite of fixed "official" prices. Real price = "official" price + correction factor (reciprocal service, bribe, additional payment, and so on). (2) Because social organization of concentration camps is not self-producing, whenever its external structural enforcement ceases it disintegrates as an entity. (3) Other, vigorously self-renewing social systems cannot be changed via structural change (short of severe physical interference): ghettos of poverty, racial distribution, free markets, or networks of criminal activity.

There are, always, social systems which have to *keep people in*, prevent them from leaving or cancelling membership, translate subtle social boundaries into their crude and visible physical embodiments. There are, always, social systems which have to *keep people out*, control and limit their entry, often resorting to similar physical embodiments of boundaries (otherwise their own self-production and existence might be endangered).

Most man-made systems are of the first kind. No spontaneous, naturally emerging system can be of the first kind—by definition. It would not, then, arise spontaneously and naturally.

J. Kenneth Benson[5] has proposed to study complex organizations by analyzing the processes through which organizational arrangements are produced and maintained. This clear focus has been mostly missing in the organization theory and

Benson's ideas represent yet another independent and parallel notion of spontaneous social orders, this time from an organizational "micro view".

Conventional organizational theory views business and governmental organizations as static structures in the state of being: it is preoccupied with man-made structural design, re-design and spatio-temporal restructuring. In our terminology, organizational theory does not deal with organizations but only with structures of a given social system. Spontaneous processes of self-organization are often labeled as "parallel", "informal", "secondary", etc., as something to be ignored or suppressed, rather than recognized as the only viable, structure-generating, self-maintaining aspect of such "organization". The word "organization" in organizational theory corresponds to our "social system"; no distinction between organization and structure is usually even attempted.

Benson starts from the right premises:[5]

> "People are continually constructing the social world. Through their interactions with each other social patterns are gradually built and eventually a set of institutional arrangements is established. Through continued interactions the arrangements previously constructed are gradually modified or replaced."

This is unmistakably an autopoietic view of the society as a process of spontaneous (not the vulgar "rational-purposeful") social construction. The product of human action then represents a constraint on human action. Yet, the social construct is *an emergent*, an outcome of mostly autonomous process. This is why attempts to regulate the process through authoritarian direction or ideological manipulation are only partly successful, whether in political empires or organizational regimes.

It is not sufficient to study social structure, "a crystallized but temporary outcome of the process of social construction".[5] There is still a fair amount of theory which takes this narrow, structural view of organization as something which can be designed in the service of specific purposes. As one can see from the radical restructuring and reshaping of organizations in the *U.S. high technology industry* (dismantling of hierarchy, flattening of the levels, *ad hoc* task forces, and emphasis on the individual): that works the best that nobody artificially imposed on people. There is no such "thing" as "formal structure" which can somehow be abstracted from the flux of the underlying organizational life. (Even if one would call such "thing" organizational morphology).

There is always the underlying social organization: a complex network of relations linking participants to each other and to the larger social world in a multiplicity of unregulated ways. It is there where one encounters the basis of power, not in the artificial "top-of-the-pyramid" scheme of things.

It is for these reasons, that Susman and Evered[22] may conclude the following about so called organizational science:

> "...the findings in our scholarly management journals are only remotely related to the real world of practicing managers and to the actual issues with which members of organizations are concerned..."

The modeling and study of the spontaneous autopoietic processes is best accomplished via computer-based simulation. One of the most important efforts is being carried out by Peter Allen and his co-workers.[1] They fully recognize that so-called "stochastic effects" are not just minor irritations and nuisance to the modeller—but the *vital force* of evolutionary change. Scenario playing and analysis are more natural and more useful to the policy maker than machine-like "predictions", "forecasts" and other extrapolation artifacts.

Another set of computer-based efforts is represented by a series of works by this author,[27-31] arguing for the *autopoietic view of social organization.*

It is more than appropriate to conclude this paper with the quotation from Lenin, a reminder to the hasty engineers of our times:

> "Small-scale commercial production is, every moment of every day, giving birth *spontaneously* to capitalism and the bourgeoisie... Wherever there is small business and freedom of trade, capitalism appears."

The above quotation demonstrates, as French historian Fernand Braudel observed, Lenin's homage and respect to the enormous creative powers of the market, exchange, self-employment and individual resourcefulness. In order to get rid of capitalism one has to root out its generating foundations: spontaneity of organization, freedom to choose, innovation and human self-reliance.

This is why *spontaneous social orders* are so important.

Omnia sponte fluant, absit violentia rebus.†

REFERENCES

1. P. M. Allen, *et al.*, "New methods for policy exploration in complex systems." In: *The Science and Praxis of Complexity*, The United Nations University, Tokyo, 1985.
2. H. Atlan, "Natural complexity and self-creation of meaning." In: *The Science and Praxis of Complexity*, The United Nations University, Tokyo, 1985.
3. S. Beer, "Preface" In: *Autopoietic Systems*, by H. R. Maturana and F. Varela, Biological Computer Laboratory, BCL Report 9.4, University of Illinois, Urbana, 1973.
4. T. G. Bergin and M. H. Fisch, *The New Science of Giambattista Vico*, Abridged Translation of the Third Edition, 1774; Cornell University Press, Ithaca, 1970.
5. J. K. Benson, "Organizations: A Dialectical View." *Admin. Sci. Quarterly*, 22, March 1977, pp. 1-21.
6. A. A. Bogdanov, *Essays in the Universal Organizational Science*, Translated by G. Gorelik, Intersystems Publishers, Seaside, California, 1980.
7. A. A. Bogdanov, *Tektologia: Vseobschaia Organizatsionnaia Nauka* (In Russian), Z. I. Grschebin Verlag, Berlin, 1922.
8. K. E. Boulding, "Foreword." In: *Autopoiesis: A Theory of Living Organization*, edited by M. Zeleny, Elsevier North-Holland, New York, 1981.

8a. K. E. Boulding, "Learning by Simplifying Complexity." In: *The Science and Praxis of Complexity*, The United Nations University, Tokyo, 1985.

9. A. Hajnal, "Systems research today: Accomplishments, conclusions, potentialities." Notes on the Cavallo Report, Hungarian Committee for Applied Systems Analysis, Budapest, September, 1983.
10. F. A. Hayek, *Kinds of Order in Society*, Studies in Social Theory No. 5, Institute for Humane Studies, Menlo Park, California, 1975.

†Motto of J. A. Comenius, "Teacher of Nations." (Let all flow in a spontaneous fashion, by force nothing can be gained.)

11. F. A. Hayek, "Notes on the evolution of systems of rules of conduct." In: *Studies in Philosophy, Politics and Economics*, by F. A. Hayek, Simon & Schuster, New York, 1967, pp. 66-81.
12. T. Hagerstrand, "Time geography. Focus on the corporeality of man, society and environment." In: *The Science and Praxis of Complexity*, The United Nations University, Tokyo, 1985.
13. C. S. Holling, "Perceiving and managing the complexity of ecological systems." In: *The Science and Praxis of Complexity*, The United Nations University, Tokyo, 1985.
14. L. Johansen, "Mechanistic and organistic analogies in economics: The place of game theory." *Kyklos*, **36**, 1983, 304-307.
15. M. Keeley, "Organizational analogy: A comparison of organismic and social contract models." *Admin. Sci. Quarterly*, **25**, June 1980, 337-362.
16. I. Kristol, "Rationalism in economics." *The Public Interest*, Special Edition, 1980, 201-218.
17. N. Luhmann, "Complexity and meaning." In: *The Science and Praxis of Complexity*, The United Nations University, Tokyo, 1985.
18. H. R. Maturana, "Man and society." In: *Autopoiesis, Communication, and Society*, edited by F. Benseler *et al*, Campus Verlag, Frankfurt, 1980, 11-31.
19. C. Menger, *Problems of Economics and Sociology*, University of Illinois Press, Urbana, 1963.
20. I. Prigogine, "Man's new dialogue with nature." In: *The Science and Praxis of Complexity*, The United Nations University, Tokyo, 1985.
21. J. Ch. Smuts, *Holism and Evolution*, Greenwood Press Publishers, Westport, Connecticut, 1973. (Originally published: Macmillan & Co., London, 1926).
22. G. I. Susman, R. D. Evered, "An assessment of the scientific merits of action research," *Admin. Sci. Quarterly*, **23**, December 1978, 583-603.
23. H. Thoben, "Mechanistic and organistic analogies in economics reconsidered." *Kyklos*, **35**, 1982, 292-305.
24. A. Touraine, *The Self-Production of Society*. The University of Chicago Press, Chicago, 1977.
25. B. Trentowski, *Stosunek Filozofii do Cybernetyki czyli sztuki rzadzenia narodom* (in Polish). J. K. Zupanski, Poznan, 1843.
26. G. Vickers, *Human Systems are Different*, Harper & Row, London, 1983.
27. M. Zeleny, "Special book reviews." *Int. J. General Systems*, **5**, 1979, 63-71.
28. M. Zeleny, "Autopoiesis: A paradigm lost?" In: *Autopoiesis, Dissipative Structures, and Spontaneous Social Orders*, edited by M. Zeleny, Westview Press, Boulder, Colorado, 1980, 3-43.
29. M. Zeleny, *Autopoiesis: A theory of living organization*, Elsevier North-Holland, New York, 1980.
30. M. Zeleny, "Organization as an organism." In: *Proceedings of the 1977 Annual Meeting of the SGSR*, SGSR, Washington, D.C., 1977, 262-270.
31. M. Zeleny, "Self-service trends in the society." In: *Applied Systems and Cybernetics*, edited by G. Lasker, Vol. 3, Pergamon Press, Elmsford, N.Y., 1981, 1405-1411.

DISCUSSION OF ZELENY'S PRESENTATION

Umpleby

In your opinion does Ashby's work on self-organizing systems and intelligence amplification address the question of how knowledge is combined in economic systems? What is missing from Ashby's theory in your opinion?

Zeleny

Ashby does not make the distinction between organization (generational) and structure (manifestational) aspects of self-producing systems.

These theories are not applicable to *social* or *human systems*. He talks of "self-structuralization" rather than "self-organization" even though he calls it self-organization. Actually Ashby later moved away from that formulation.

Umpleby

Can you give a reference?

Zeleny

Compare his *Design for a Brain*, (1960) and his "Principles of the Self-Organizing System" (1962). Ashby stated that the description of living systems in terms of state-determined behavior is adequate and the term "self-organizing system" could be dropped from use.

Sadovsky

Give us please an example of a social system which has man-made and spontaneous components. Is it possible to say that spontaneous components are unpredictable results of man's activity or in other words that the elements and subsystems, of a spontaneous system are man-made, but the spontaneous system itself is not man-made?

Zeleny

Examples of spontaneous systems: markets, money, firms, families, clubs, towns, ideologies, interest rates, etc. Anything having boundaries to keep people out rather than to keep people in.

Examples of man-made systems: military formations, concentration camps, managerial hierarchies, command economies, artificial towns and social institutions, etc.

Spontaneous orders are *unintended* (unplanned for) emergents of human interactions. These spontaneous outcomes are unpredictable only to the extent of our ignorance of the *generating* mechanisms.

Von Glasersfeld

The distinction between spontaneous systems which we do *not* understand and man-made systems which we do understand is relevant to politics. Politicians (of every color) often try to make people believe that the particular man-made system they represent is *the* spontaneous system that everyone wants but has not yet realized. This, surely, is one root of demagogery. A fully "spontaneous" system could arise only through what I would call "collaborative dialectic."

Umpleby

Are the terms centralization and decentralization a suitable paraphrase for your terms man-made and spontaneous organization? Governments can be thought of as oscillating between centralized and decentralized policies. In the U.S. the Democratic Party tends to suggest centralized policies while the Republican Party prefers decentralized policies. Even in one party systems one can observe such changes. For example, in China centralized policies were pursued under Mao, but lately there have been experiments in decentralization.

Zeleny

Centralized systems are *typically* man-made by definition of the central designer. Spontaneous ones are *typically* non-centralized *but* there could be spontaneous transformation of decentralized into centralized, and vice versa.

The policy oscillations I would attribute to our poor understanding of the proper roles design and self-production play in social systems.

Mitroff

Why did early systems ideas not catch on?

Zeleny

The nature of scientific revolutions.

Lapin

You strongly stated that spontaneous systems are more viable with respect to external impacts than consciously created systems. Is this an intuitive statement, or is there some scientific proof for it?

Zeleny

I didn't mean to state that *all* spontaneous orders are more resilient to external disturbances. Some are very brittle and can be destroyed (a free democratic society). Some man-made systems can be "shielded" from external disturbances by man-made "barriers"; thus they can survive longer. *Removing the barriers*, spontaneous systems are better equipped to survive.

Look at a system's boundary and watch which way people are crossing it. People do not leave a spontaneously self-organized system.

CHAPTER

10

THE USE OF THE THEORY OF ALGORITHMS IN GENERAL SYSTEMS THEORY

Igor A. Lavrov

General systems theory is a relatively new science. Its principal aim is analysis and formalization of the basic systems notions as well as creation of its own specific methods of investigation. Thus, it is only natural that all kinds of methods and notions of different directions of modern mathematics such as algebraic and topological conceptions, stability theory and category theory, differential equations and functional analysis, control theory and many others find application within this field. Some directions of mathematical logic and the theory of algorithms have also made their contribution to general systems theory. Today these sciences are certainly on the rise. New methods are being created, old problems are being solved. Many domains of mathematics borrow ideas from mathematical logic and the theory of algorithms. The explanation of this fact is that many notions of these sciences such as language, axioms, conclusion, provability, computability, complexity and some others are closely related to the basics of mathematics.

The aim of this report is to familiarize specialists in general systems theory with a number of important directions in the modern theory of algorithms. My preferences, as well as the information I possess, predetermined the choice of these directions. But I also hope that these methods and ideas will be useful for specialists in general systems theory. I need not adjust my account under the canons and standards of general systems theory. I shall deal only with those objects that can be related to the general mathematical culture. A specialist in general systems theory can easily modify the ideas of this report and use them in his own field.

It should be mentioned that such a close interrelation of one science with another creates two important tasks:

1. The need to use the well-developed methods of the theory of algorithms in general systems theory;

2. The need for the theory of algorithms to solve the problems of general systems theory.

In this report I shall dwell on two directions which seem to me to be important: complexity and the theory of numberings.

1 COMPLEXITY

The fundamental computing tool of the theory of algorithms is the idea of a *Turing machine*. Specialists in general systems theory are quite familiar with this idea so there is no need for me to give the exact definition here. It should be remembered that all Turing machines can be *effectively numbered* by natural numbers $N = \{0,1,2,...\}$:

$$M_0, M_1, M_2, \cdots$$

The details of that numbering are immaterial. But this numbering possesses the following important properties:

1. If we know some concrete Turing machine M, we can find effectively (in finite time) a number a, the index of the machine M, so that $M = M_a$;

2. If we take any Turing machine M_a, we can restore effectively (in finite time) its basic parameters by the index a: symbols, states and the program of operation. The fulfillment of these conditions guarantees specific invariance of the theory of algorithms.

The complexity of a Turing machine can be measured in different ways. The most popular evaluation of the complexity of a Turing machine was suggested by C. Shannon: $n \times m$ where n is the number of symbols and m is the number of active states of the machine M. Here there is one problem which remains unsolved. Under what minimum $n \times m$ does the universal Turing machine exist? A number of authors have reduced the problem to the following:

1. There exists a universal Turing machine of size 4 x 6;

2. There are no universal Turing machines among the machines of size 3 x 2;

3. There is a small number of Turing machines left which have not been investigated for universality.

With each Turing machine M_x we connect a one-place partial recursive function ψ_x which is computed on M_x at some fixed definition of computation. Later on we restrict ourselves to one-place functions, since the transfer from many-place functions to one-place functions offers no difficulty, but all typical details of computations take place in that case as well. The functions ψ_x are called *partial recursive functions* and they play the fundamental role in the theory of algorithms. According to Church's thesis, the class of partial recursive functions coincides with the class of intuitive computable functions. If ψ_x is a total function, then it is called a *recursive function.*

In the sequence $\{\psi_x\}$ each concrete partial recursive function occurs not once but an infinite number of times, since each function has different ways of computation. But the indicated sequence possesses a series of more important merits. For example:

1. Any other sequence of partial recursive functions (not necessarily all), which possesses effective properties following from the similar properties of the Turing machine numberings mentioned above, can be reduced by constructive methods to the given sequence $\{\psi_x\}$.

2. The phenomenon of a fixed point holds for the sequence $\{\psi_x\}$: for each recursive function α there exists a number e such that $\psi_{\alpha(e)}(x) = \psi_e(x)$. It should be mentioned that the class of recursive functions does not possess similar numberings. Analogously, if we try to arrange all partial recursive functions in sequence without repetitions (the way it can be done was demonstrated by R. Friedberg), then a very complex numbering is obtained and some useful properties are lost.

The attempt to find an adequate definition for the complexity of computation of the partial recursive function ψ_x, taking into account different numerical parameters of machine M_x and size measures of the computation process, meets some objective difficulties. It is well known that when using simple suppositions imposed on the notion of complexity (axioms of Blum) there is the effect of "speeding up computation". Naturally, in such a case, when defining the complexity of computations, we have to resort to some minimum parameters (e.g. minimum program, minimum time for computation and so on). This is not always convenient for practical purposes.

In the classical theory of algorithms E. Post introduced some variants of comparisons of computational complexity in the form of reducing some objects to others. Now we are passing over to a short review of ideas and results in this direction. Later on we shall consider only proper (not ϕ and N) subsets of natural number set N.

We say that a set A is reducible to a set B if there is a recursive function f such that for all x

$$x \in A \leftrightarrow f(x) \in B$$

It is not difficult to see that reducing function f in many respects resembles such an object in theoretical programming as the translator. Abbreviate A reducible to B as $A \leq B$. The sets A and B are equivalent (A ≡ B) if $A \leq B$ and $B \leq A$. From this point of view equivalent sets are indistinguishable and the family of all sets is decomposed on the classes of equivalent sets. These classes are used as a measure of complexity of the sets. They make it possible to consider some hierarchical conceptions and frequently determine some fine properties of the sets. We shall denote the set of classes of equivalent sets, called *degrees*, as L.

Among elements of L there exists the least element which we will denote by 0. It contains only the so-called *recursive* or *decidable* sets. The partial ordered by ≤ set L is the upper semilattice with a rather complex algebraic structure, the description of which was given by J. Ershov. In the class of distributive upper semilattices L possesses a rather high power of universality.

The notion of *recursively enumerable set* is one more fundamental conception of the theory of algorithms together with the Turing machines and partial recursive and recursive functions. Such are the ranges of values of partial recursive functions (we can give some other equivalent definitions). The range of values of the function ψ_x will be denoted W_x. Thus we have the effective numbering of all recursively enumerable sets $\{W_x\}$. This numbering possesses the same useful properties as the numbering $\{\psi_x\}$. The upper semilattice $L_{r.e.}$ of classes of equivalent recursively enumerable sets has been studied most intensively. In $L_{r.e.}$ besides the least element 0 there exists the largest element 1 which can be represented by the so-called *creative sets*. Thus each creative set has maximum complexity among all recursively enumerable sets. The creative sets are met frequently in some branches of mathematics. For example, in the suitable numbering of logical formulas of first-order arithmetics, the set of indices of provable formulas is a creative set. From this fact and properties of creative sets we can prove very easily the famous Gödel theorem about incompleteness of formal first-order arithmetics of Peano.

In addition to recursive and creative sets there is a denumerable set of recursively enumerable sets, and their degrees comprise a rather whimsical algebraic structure. Many of these degrees are *incomparable*: a $\nleq$ b and b $\nleq$ a. The algebraic description of $L_{r.e.}$ was given by S. Denisov. His description is rather complicated. If we denote $L_{r.e.}$ (a) the upper semilattice of degree b such that $a \leq b$ then for all $a \neq 1$ semilattices $L_{r.e.}$ and $L_{r.e.}$ (a) are isomorphic:

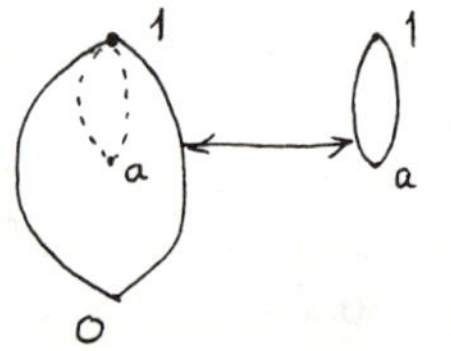

This method of defining a measure of complexity of mass problems has significant advantages over the direct numerical evaluation of computing parameters. Note that E. Post also introduced some other kinds of reducibility of sets, which took greater account of the character of the mathematical approach. Most known from that reducibility is *Turing reducibility* $\leq_T$. Classes of Turing equivalent sets are called *degrees* of *unsolvability*. The majority of articles on the theory of algorithms are devoted to the study of degrees of unsolvability. In this direction Friedberg and Muchnik created a specific and highly effective "*priority method*" which, as we know, has not been yet used in other domains of mathematics.

2 THEORY OF NUMBERINGS

The idea of numbering objects of nonnumerical nature by natural numbers and transportation of statements about these objects into formal first-order arithmetics was first used by Gödel when he proved his remarkable theorem about the incompleteness of Peano arithmetics. A. N. Kolmogorov called for a systematic study of the numberings of arbitrary sets. W. A. Uspenski was occupied with the realization of this idea. In various countries specialists investigated the theory of numberings (J. Myhill, R. Friedberg, J. Dekker, H. Rice, D. Lacombe, A. Lachlan, R. Soare, C. Yates and so on). The principal discoveries in the theory of numberings were made by A. Malcev and J. Ershov.

Let A be an arbitrary denumerable or finite set of objects of any nature. A *numbering* of A is any mapping ν from the set N onto A:

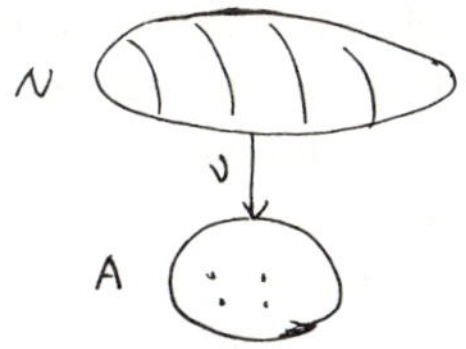

The pair $\Pi = \langle A, \nu \rangle$ is called a *numbered set*. If $\nu n = a$, then n is an *index* of object a at the numbering ν. We can consider the numbering ν as a distinctive coding of the objects of set A by natural numbers. For the given numbering ν the set N is decomposed on the classes containing natural numbers which are indicies of one and the same object of set A.

Non-trivial set A possesses many different numberings. This makes us compare different numberings of set A. A numbering ν_1 of set A is *reducible* to a numbering ν_2 of set A, if there exists a recursive function $f(x)$ such that for all $n \in N$

$$\nu_1(n) = \nu_2(f(n)).$$

In other words if we know any index of an object a at numbering ν_1 then we can effectively find (by function f) some index of the same object a at numbering ν_2. Abbreviate ν_1 reducible to ν_2 as $\nu_1 \leq \nu_2$. The numberings ν_1 and ν_2 are called *equivalent* if $\nu_1 \leq \nu_2$ and $\nu_2 \leq \nu_1$. From the point of view of the theory of numberings, equivalent numberings are indistinguishable. The family of classes of equivalent numberings L(A) of set A are a partial ordering set by ≤ and can be used as a measure of complexity of set A. It should be taken into account that if set A consists only of two objects a and b, then the numbering ν defines some set T (indicies of object a), and the reducibility of numberings of set A models the above mentioned reducibility of sets of natural numbers. Thus it is a special case of the general theory of numberings.

The partial ordered set L(A) has a rather complex structure even for the very simple set A. So specialists in the theory of algorithms having used *incomparable elements*, that is such numberings that $\nu_1 \not\leq \nu_2$ and $\nu_2 \not\leq \nu_1$. This means that for some function there exist programs of its computation which cannot be translated one into the other.

A number of problems of the ordinary theory of algorithms can be interpreted in the theory of numberings by a peculiar method which leads to some new important notions. So let us consider the classic problem of extension of a function: for the given partial recursive function $\psi(x)$ try to find some recursive function $f(x)$ such that $f(x)=\psi(x)$ where $\psi(x)$ is defined. This problem is unsolvable in the general case. There exist partial recursive functions which cannot be extended to recursive ones. Passing over to the theory of numberings we come to the notion of precomplete numbering introduced by A. Malcev. The numbering ν is called *precomplete* if for any partial recursive function $\psi(x)$ there exists a recursive function $f(x)$ such that $\nu f(x)=i\psi(x)$ where $\psi(x)$ is defined. A precomplete numbering is called complete if at points where $\psi(x)$ is not defined the function $f(x)$ takes as its values the indicies of some special element $a \in A$. The notions of complete and precomplete numberings play the fundamental role in the theory of numberings. The existence of complete or precomplete numbering of set A testifies to a strong universality and the importance of class object A. Only complete and precomplete numberings have such important properties as a recursion theorem, a fixed-point theorem, an effective fixed-point theorem and so on. The above indicated numberings of partial recursive functions $\{\psi_x\}$ and recursively enumerable sets $\{W_x\}$ are complete numberings: their special elements are an everywhere indefinite function and an empty set.

If A is a family of recursively enumerable sets or partial recursive functions, then among the numberings of A we can select the so-called *computable numberings* (for example, in the first case it means that the relation $x \in \nu y$ is recursively enumerable itself). The family of classes of equivalent computable numberings of set A denote L°(A) it can be partially ordered by the relation $\leq$. If at L°(A) there exists the largest element a, then the numberings from a are called *main computable numberings* of A. There is a trivial analogy here with the notion of a universal machine. Any program of computation for any machine can be reduced to a program of the universal machine.

Comparing numberings of two sets A and B, the basic notion is that of morphism. A *morphism* μ from numbered set $\sigma = \langle A, \nu_1 \rangle$ to numbered set $\xi = \langle B, \nu_2 \rangle$ is a mapping from A to B for which there exists a recursive function f(x) such that for all $n \in N$: $\mu\nu_1(n) = \nu_2(f(n))$.

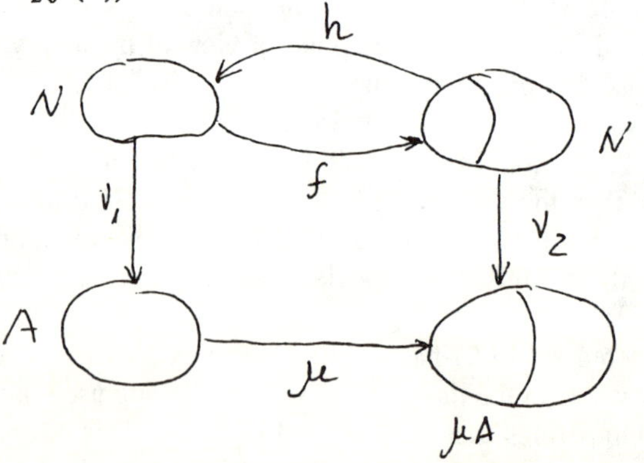

In other words, if we know any index of object $a \in A$ in numbering ν_1, then we can effectively find (by function f) some index of image μa in numbering ν_2. So the function f models the mapping μ on the indicies. The set of all morphisms from σ to $\mathcal{L}$ is denoted $Mor(\sigma,\mathcal{L})$. The investigation of $Mor(\sigma,\mathcal{L})$ and the allotment of $Mor(\sigma,\mathcal{L})$ by "good" numberings in particular are the central aims of this research in the theory of numberings. Let us consider one more classic problem—the problem of inversion of function: for the given recursive function $f(x)$ try to find some partial recursive function $h(x)$ such that

$$fh(x) = x$$

where the left side is defined.

Passing over to the theory of numberings, we come to the following: also for the given recursive function $f(x)$ try to find some partial recursive function $h(x)$ such that the indicies of images μA in numbering ν_2 satisfied the following:

$$\nu_2 fh(x) = \nu_2(x).$$

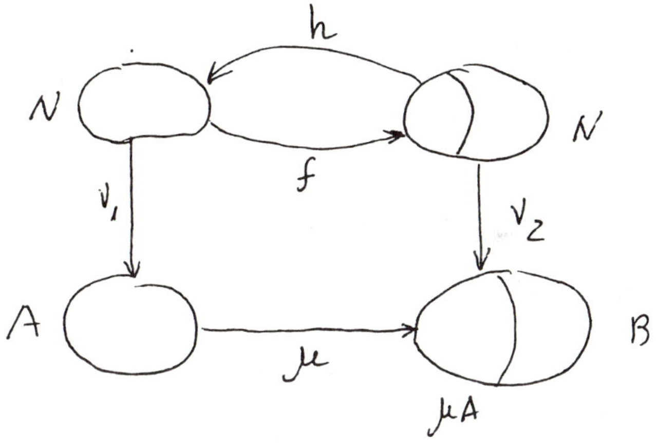

The choice of this inverse function h with some additional properties, which are connected with the behavior of h on indicies of objects from $b - \mu A$, will determine this or that kind of *subobject*. The case of *standard subobject* (Lachlan's class), where $h(x)$ is a recursive function, is of the greatest interest. The standard subobject is closely connected with complete numberings. Among the important unsolved problems is the problem of satisfactory description of any standard subobject. This description could clarify many problems of the theory of algorithms. For some other kinds of subobjects, which are close to standard subobjects, there are exhaustive descriptions.

At present the theory of numberings is rapidly developing. Using its methods we can have a new look at such directions as for example: computable functionals, constructive algebras, and so on. Without doubt a specific but rather general approach to problems of the theory of algorithms can play an important role in other sciences.

My final slogan is: to number and reduce, to number and reduce, and once more to number and reduce.

DISCUSSION OF LAVROV'S PRESENTATION

Klir

Computational complexity has usually been studied in terms of the worst cases in individual problem classes. Recently, interest is shifting in the direction of studying "average" or "typical" cases. Has some research in this new area been done in the Soviet Union?

Lavrov

Yes, if I understood you correctly. Some assumptions gave only a general picture of process. In these cases we can study a part of a problem, and this part could have a lower level of complexity than the whole case.

In the USSR Guravrev's group takes a similar approach. (Computer Center of the Academy of Science, Moscow).

Umpleby

How does the work in mathematics help us to understand some of the problems we have been discussing—paradox, implementation, and the formulation of morally important problems to work on?

Lavrov

I try to answer this question in my paper. We can also look at the history of mathematics, especially in the 20th century.

DISCUSSION OF METHODOLOGY

Umpleby

It seems to me that at least two views regarding paradox have been used in our discussions so far.

1) If a paradox arises at a particular level of analysis, perhaps it can be resolved by focusing on the level of ultimate purpose.

2) If paradox results from the relationship between the observer and the social phenomenon observed, difficulties may be more persistent.

Mitroff

The purpose of paradox is to show how limited our models are. Do not play a theoretical game with the term "paradox."

Allyn

To keep the peace, prepare for war. The way to do is to be.

Mitroff

Pointing out paradoxes is a way to shake up policy makers. Paradox is a method not a logical difficulty.

Umpleby

Ian's statements just now have greatly clarified for me his position on the subject of paradox. Perhaps we should now consider what problems can be dealt with by the methods presented by Kossov, Klir and Lavrov and what problems cannot be dealt with by these methods?

Klir

Reconstructability analysis can be used for dealing with problems in ecology.

Umpleby

Perhaps problems concerning the allocation of physical resources lend themselves to mathematical models whereas problems of goal formulation and conceptualization require linguistic and communication methods.

Apokin

Deterministic methods are not appropriate for social systems.

Zeleny

It is not an issue of mathematics or non-mathematics, deterministic models or non-deterministic models, rather what is appropriate to the issue addressed.

If social systems are not machines, then certain methods based on mechanistic assumptions will not work.

Mitroff

University departments are not suited to dealing with global problems.

Zeleny

We must first evolve the methods we need. Research on these methods is not being funded.

Ben-Eli

The methods we already have are not being used where they can be most successful.

Zeleny

In economics there is a debate over whether economic systems are machines or organisms.

Von Glasersfeld

What distinctions are made in the USSR regarding organisms and machines?

Sadovsky

Biological and social organizations are self-productive.

Rapoport

A social system is a special type of system. People have the ability to formulate goals. We need to have a better understanding of paradox.

Lapin

There is a view in the West that Marxism is unified. This is not true. We need to use a whole set of methods, to view the world as multileveled. We cannot reduce social to biological, biological to chemical. It is necessary to have qualitatively new methods for the social sciences. Not just for training. Also experiments—dozens or hundreds of field trials. There is of course the problem of coordinating the use of the results of the experiments.

Pospelov

It seems to me that the issue of the mathematics of the non-living and the living has been posed incorrectly. Conventional mathematics can successfully describe the work of the cardiac muscle or the system of motor neurons which move the legs even though the heart and the legs are biological objects. But this same conventional mathematics does not have the instruments for describing geological processes, even though geology deals with inanimate objects. The situation in chemistry is analogous. This means that it is not a question of creating a mathematics of the living organism, but rather of developing the needed theories of biology, physiology and psychology (also of geology).

Von Glasersfeld

Would the Soviet participants please explain what dialectics currently means in the Soviet Union.

Sadovsky

Ernst Von Glasersfeld's question reminds me that Compte is supposed to have once asked Hegel to describe his philosophy briefly, understandably, and in French. Hegel's answer was negative.

Dialectics is not just a concept. It is close to a world outlook. Dialectics is a special kind of philosophical investigation. The purpose is to establish some general rules of development—laws which describe development. The essence of such laws is to establish different kinds of contradictions. Paradox, as we have been using the term, is close to dialectics. Some people consider dialectics to be a method as well as a logic. I think there is no special dialectical logic. Dialectics mostly has looked at parts and wholes in a philosophical sense.

Lapin

In addition to what Sadovsky said, I would like to stress the importance of understanding how Marx became a dialectician. This was the subject of a book I wrote, *The Young Marx*.

Currently there are a variety of views among Marxists on what dialectics is. Dialectics is, from my point of view, a general theory of development. The basic element of it is contradiction. Paradox can be considered as one source of development. Contradictions are a source of development in socialist as well as capitalist society. A basic concept for Marx was that the whole is greater than the sum of the parts. Systems qualities of social systems are internally contradictory. Throughout the history of human society humanity has developed greater and greater freedom. The heterogeneity of the world is manifested in the most complex way, proceeding from abstract to concrete. The abstract is one aspect of reality. Of course, Sadovsky and I have given you only very short expositions of some important aspects of dialectics.

CHAPTER

11

THE METHODOLOGICAL UNITY OF SYSTEMS AND CONTINGENCY APPROACHES TO THE ANALYSIS OF MANAGEMENT SYSTEMS

Vladimir S. Rapoport

1 INTRODUCTION

It is universally recognized that the management of social systems is one of the most demanding areas of systems research applications. The use of the methodological concepts and principles of the systems approach in analyzing the management systems of business organizations and other human activity systems made it possible, apart from producing practical results, to extend and develop the theoretical concepts of the subject and methods of studying the most complex phenomena of social development. A huge number of papers and books has appeared during the last quarter of this century testifying to the great effectiveness of systems research into management techniques, forms, and processes.

The wide-ranging application of the systems approach to analyzing management problems in social systems and the considerable accomplishments in organizational development theory were conducive to identifying new methodological problems.

In the first place, it became clear that systems analysis of organizations, their behavior and management, while contributing to understanding some general relations and regularities, does not yet provide opportunities for a strict scientific choice of concrete solutions to quite specific problems arising in a wide variety of business situations.

Thus far, the attempts to develop the methods of decision choice and substantiation based on sufficiently explicit and objective criteria, standard and practitioner-oriented procedures, norms, and calculation techniques have failed.

The major obstacle is that the assumptions, models and principles used in general systems theory are too universal. In the general form they hold for all types of systems and for any state thereof. A particular case study, however, includes many variables, and their relationships are not subject to operational definition. The point is that a strictly systems approach to the study of organizational behavior requires models that are too complex for practical implementation.

No considerable advances have been made in organization theory either. Systems analysis is based on abstractions from the descriptive characteristics of systems which are very different—economics, technology, information, etc. Therefore, the theoretical framework and methodological tools of the various scientific disciplines are not conducive to developing a sufficiently complete knowledge about the organization as a unique system nor to determining the regularities of its evolution. Effective here may turn out to be only an interdisciplinary approach and identification of a concrete subject matter of research adequate to the phenomena under study.

Accordingly, many consider the contingency approach to be an alternative to the systems approach in the study of organization and management systems. Since the 1960s independent research based on the contingency approach has been conducted concurrently with systems research, and during the 1970-80s it even prevailed as indicated by the number of publications. Even a "General Contingency Theory" was claimed to evolve (American theories of management, 1978; B. Milner, 1982).

There are considerable methodological differences between the two approaches, at least given the present state-of-the-art. In brief, it may be noted that the major criterion of organizational effectiveness, from the systems approach standpoint, is wholeness while from the contingency approach point of view, it is an opportunity for the fullest possible manifestation of some basic relationships among the system's components.

The published results of research based on the contingency approach indicate, however, that while producing quite satisfactory partial results, it does not ensure a sufficiently complete solution of problems.

Thus, there are two different approaches to analyzing organizations, each one having an objective scientific value but neither of them satisfying the users. Of course, both the systems and contingency approaches are based on certain objective characteristics of management systems and regularities in their development but not to the extent of securing a valid choice for organizational solutions. Hence, there arises a pressing scientific problem: to find out whether an integration of the two approaches is feasible. If yes, then on what basis, and will it suffice for the accomplishment of the aforementioned objective?

2 THE SYSTEMS APPROACH AND THE SCOPE OF ITS APPLICATION

The systems approach to the analysis of organizations has been exercised on a sufficiently developed theoretical and methodological base both in the study of individual components of management systems by the respective disciplines and in the comprehensive study of organizations on the basis of multidisciplinary research integrated by cybernetics and/or sociology. (Gvishiani, 1972; Litterer, 1974).

Back at the "presystems" stage of organizational research a number of notions and ideas evolved which are still scientifically valid. These ideas play a central role in organization theory. In the first place, these are: segregation and description of management as a specific component of organization playing a special role and exercising unique functions; accumulation of knowledge about management as a process of decision making and implementation which determines a universal structure for the process; defining, dividing, and coordinating in order to form operational notions about management relationships and communications; identification of the line and staff relations within an organizational structure.

The systems approach has introduced new notions and criteria that have helped to develop and extend established knowledge, to give new interpretations of known phenomena, to bring order to methodological tools, and to identify new regularities and principles.

The concepts in the systems approach which have most influenced the development of organization theory are the notions of wholeness and hierarchy in organizational systems and the idea of an organization as an open, multi-objective system. Consider briefly, what inferences and methodological problems result from the application of the above statements.

As the *principle of wholeness* suggests, the properties of the organization as a system cannot be described and defined as a simple sum of the properties of its individual components. A goal-seeking interaction of system components determines their behavior (which thus becomes interrelated) in such a way so that some specific result is produced inherent only in the concrete system.

A knowledge of the properties of individual elements and regularities of their independent functioning and development acquired within the frameworks of traditional special disciplines (economics, technology, sociology, etc.) is therefore a necessary but insufficient prerequisite for analyzing a system as a whole. Just as important is the knowledge of laws regulating the interaction of the heterogeneous components of human activity systems striving toward specific objectives within particular environments. Of interest here are not the properties of components but the property-related behavior necessitated by goal-seeking activities.

The most important feature of organizations as social systems is that all components thereof are classified into two types—active and passive (Burkov, 1977; Rapoport, 1979, 1982).

Active elements have their own goals and interests, hence they considerably affect the selection of goals for the system and its subsystems and they are capable of introducing changes into their functional orientation according to the tasks set. Such properties are possessed by all social elements of the organization—individual employees, groups thereof, and department staffs viewed as independent entities. Passive elements are all the rest of the social system components, such as the implements, means, or conditions of work having no independent functions in decision formulation and implementation.

The functions of passive components are determined by their characteristics and the technology being applied. Complex though the composition and patterns of the elements' relationships may be, the relations between their inputs and outputs, if each element is viewed as a system, are strictly objective. They are subject to the well-known laws of nature and lend themselves to simulation by standard methods.

No conclusion of this kind can be made with respect to the behavior of active components. The laws of historical materialism, political economy, and sociology which determine the dependence of human behavior on motivation and the external environment belong to rather large sets. In regard to concrete subjects in their everyday behavior, these laws manifest themselves in deviations from the latter and the subjective perception of the accomplished results (Afanasyev, 1973).

It must also be kept in mind that the operational goals of organizations and the business environment vary so much that they lend themselves to systematization only in the most general form. Much more varied is the interorganizational behavior of separate active elements related to the accomplishment of objectives. Given a huge number of behavioral alternatives for each element and, respectively, a much larger number of combinations thereof, the simulation of organizational behavior becomes rather difficult.

Keeping in mind that the choice of a behavioral alternative is, in each concrete case, largely determined by the subjective interests and criteria of the decision maker, the input-output relations in the functioning of active system components are always of a probabilistic nature. This imposes additional constraints on organizational behavior simulation.

These properties of human activity systems inhibit a formal quantitative analysis of organizations. Also, highly bureaucratic, directive management of organizations lacks scientific substantiation, since the feasibility of securing the wholeness effect is rather probabilistic. Hence, the theoretical framework of a bureaucratic organization and the relevant concepts (Haynes and Massie, 1969) have to be substantially revised.

Systems analysis shows, however, that the relatively high autonomy and subjective orientation of active elements reduce the opportunities for a formal centralized control of social system activities, thereby enhancing the accomplishment of the wholeness effect through their *self-organization*. A high adaptability of each component to the influences of other components (and not only of a managerial entity) compensates the shortage of directive information (within a certain range of allowed deviations) and secures the required wholeness of the system in the process of its functioning. This property makes it possible to considerably simplify the organizational behavior model by discarding many communications amenable to self-regulation.

The systems approach to interpreting organizational *hierarchy* has also contributed to the revision of established knowledge and concepts. Whereas the traditional concepts identified hierarchy primarily with subordination, with a direct and full dependence of the lower organizational levels on superior ones, systems analysis exposes more diverse relationships.

The interrelationship between the behavior and properties of system elements manifests itself in so far as its general goals are commensurate with both the capacities and interests of the active elements making up part of the system. Therefore, apart from the direct dependence of each subsystem's behavior on its place and role in the system, there is also an inverse relation between the system goals and structure and the properties and interests of its elements.

The aforementioned effects take place not only in the initial stage of goal setting but also, and just as strongly, during decision implementation when goals are adjusted to take account of the response of performers and the accomplished results. If these facts are not taken into consideration when constructing an organizational behavior model, then the most important factors determining the organization's effectiveness will be missed.

Of considerable significance in the analysis of hierarchical organization structure is its multiobjective nature. The fact that there can be several top level goals of equal priority for the system as a whole makes one view the concept of hierarchical structure quite differently. It is quite possible to view the organization as a family of single-target subsystems (program, project structures), multiobjective (functional, divisional) structures and a combination of both (matrix structures). In any case, the effect of system wholeness is achieved not through a mere hierarchical integration (or decomposition) of goals (Milner et al, 1983).

One of the most important concepts of the systems approach, drastically changing traditional knowledge about the regularities of organizational behavior, is their interpretation as *open systems*. The external environment that the open system has to interact with often causes frequent changes in the operating conditions and objectives of the organization. Accordingly, this requires respective alterations to its basic characteristics, or building in ones that would provide for the maximum flexibility of organizational behavior, thereby allowing the organization to adapt to the changing demands and conditions. In any case, it becomes so difficult to develop, in advance, a model of organizational behavior in the changing environment that many researchers view the problem as unresolvable. The problem is aggravated by the environmental uncertainty which does not permit predicting its variation with the desired accuracy (Meyer et al., 1978).

Thus, it is quite evident that the systems approach extends and substantially changes earlier knowledge about the content and nature of relationships within organizations, their purposes and methods of management. As is well known, the systems approach as such is just a methodological tool for ordering and developing scientific knowledge which must be applied in order to formulate concrete theories and concepts. Thus far, however, the systems approach as applied to organization theory has succeeded, first, in proving the inadequacy of previous models of organizational functioning and development and, second, in discovering far more complex relations and structures.

The next phase of systems research must aim at defining sufficiently strict qualitative or quantitative criteria in order to make it possible to evaluate the rationality of concrete system structure and behavior in a specific environment and to select operational decisions by the variations in individual organization parameters, thus ensuring the maximum correspondence of their characteristics with the objective requirements.

3 FORMS OF MANAGEMENT ORGANIZATION AND SPECIFICS OF THEIR ANALYSIS

The major difficulty in analyzing and simulating human activity systems is a great variety of behavior patterns displayed by the active elements. The choice of behaviors is largely influenced by subjective factors. Experience has shown that the most effective means of purposefully ordering and regulating system behavior is by bureaucratizing relations among the elements. Since technological and organizational relationships lend themselves readily to bureaucratizing, the organizational structures become the most important characteristics of human activity systems.

Analyzing and setting up technical production structures are easy from a methodological point of view, since they reflect the pattern of relationships among the passive elements of the system. At present, there are rather strict criteria and sophisticated methodological tools for developing and setting up an organizational structure.

Systems analysis indicates that management organization is far more complex than the operating one. It mirrors the relationships of the active elements of the system dominated by the operational goals, individual and group interests, subjective capabilities and the orientations of decision makers and operators.

The formal management structure, in its generally accepted interpretation, regulates only the subordination relationships, i.e. the "command-subordination" relationships. In this connection, it was generally considered as directly related to the production structure, and in many papers both structures were actually identified (Drucker, 1974; Ovsiyevich, 1979).

The growing scale and complexity of business activities, their multiobjective nature, changes in goals and the conditions of their implementation have resulted in the fact that along with the tendency toward division and specialization, management is increasingly dominated by a trend toward the integration of managerial components of organizationally independent subsystems when solving problems of systemwide significance. Behind the integration is a joint responsibility for the accomplishment of common goals and joint decision making. Accordingly, as was mentioned above, there evolve new management relationships qualitatively different from the reporting relationships in hierarchical systems.

What is more, since the dynamic and uncertain environment rules out the bureaucratic organization, a conclusion suggests itself that the traditional management structure, laying out only the pattern of subordination relationships, cannot adequately regulate organizational behavior and managerial processes.

Hence, the organization structure as a characteristic of a human activity system may be viewed only as a prerequisite for its functioning, having no direct linkage with the accomplished results. This relationship may be formulated as follows: the lack of a structure consistent with the established environment and goals makes them unrealizable; there are, however, different structures meeting the requirements set.

Should the thesis be true, important theoretical and practical questions arise: to what extent do the different types of structure affect organizational behavior and effectiveness; is there an optimal structure and what factors determine its selection? It is possible to answer these questions only by extending the concept of structure as a pattern of different types of relationships—vertical (subordination) and horizontal (relations of joint decision making or interaction).

If, however, the vertical relations are naturally regulated by the line subordination schemes, then the description and regulation of horizontal relations require the installation and control of interaction procedures during collective decision making—the procedure of development and discussion of alternatives, their coordination, approval, and documentation. Accordingly, there are normative documents and regulations—descriptions of positions, departments, program and project management systems, control procedures, record-keeping functions, etc.

Experience shows that different though the forms of joint decision making may be, the fundamental principle of management organizations is that of unity of command based on the unity of direction and control in accomplishing concrete goals. Therefore, every structure of horizontal relationships must observe a certain subordination ensuring a single level of decision making and responsibility for the ultimate result. The distinguishing characteristic of horizontal relationships is that this level must not necessarily be the top one in the vertical hierarchy (with respect to other participants in joint decision making).

The analysis of the evolution of horizontal relationships (program, project structure) and methods of their combination in matrices with the traditional line-hierarchy forms shows that the top level of organization management acts as an integrating link for the entire management system. This expands the span of top management control still further and makes it necessary to look for ways to decentralize decision making while preserving general control of organizational behavior, ensuring its maximum effectiveness.

As has been shown (Burns and Stalker, 1961; Galbraith, 1973) most adequate to problem solving is an organic behavioral model and the installation of respective managerial relationships. The model, fully accounting for the subjective orientation of the system's active elements, places the major emphasis on the following factors determining the behavior of each element and system as a whole. Assignment of a strict and undivided responsibility for the accomplishment of concrete results ensuing from the organizational goals; installation of an effective incentive scheme to improve operations efficiency; providing for consistency between the skills and qualities of managers at all levels and the nature of problems solved.

This may lead to the formulation of basic guidelines for the analysis and improvement of managerial relationships in business organizations:

- decomposition and structuring of system goals ensuring, on the one hand, its wholeness and proper assignment of undivided responsibility for the accomplishment of subgoals and objectives between managerial levels, on the other;

- ensuring commensurability between the administrative responsibility for the accomplishment of the assigned tasks and material incentives for improved performance;
- ensuring correspondence between the assigned functions and individual traits of managerial personnel.

It must be kept in mind that organizational structure as a characteristic of a management system consists of two kinds of components—invariant components and variable components (Rapoport, 1979).

So, a concrete system operating in a certain environment, has several optimal structural parameters stipulated by the regularities of organizational functioning and development closely related to the parameters of the technological base and the content and nature of environment relations. This is, for example, the division of the system into organizationally independent subsystems, the order of their subordination, the number of hierarchical levels, the distribution of goals and subgoals between organizational units, and the coordination procedure for functional block (parts thereof) interaction. The above parameters determine the stable organizational relationships which are invariant within a certain range of functioning conditions. Naturally, they have the greatest impact on the behavior of the organization as a whole and its effectiveness.

The variable component of organizational structure is primarily the content and pattern of communications among the management system components within the frameworks of established relationships. The variable parameters reflect the impact of the specific environment of managerial activities including the subjective factors and informal relationships. They are subject to control at different management levels and at different time intervals. In fact, the wider the range of variable parameter changes, the better the adaptation capabilities of the organization retaining the same level of effectiveness.

As follows from the previous analysis, the primary purpose of organizational analysis is to determine the interrelationships between the technological and environmental variables, on the one hand, and the invariable parameters of organization structure, on the other, providing for the greatest organizational effectiveness. The solution of this problem is behind the majority of concepts of the contingency approach.

4 BASIC PRINCIPLES AND RESULTS OF APPLYING THE CONTINGENCY APPROACH

It is hardly possible to find in the papers of contingency approach proponents any universal formula adequately reflecting its methodological essence or some general principles. The most typical definitions, made at the early stages in the development of the approach, were the ones suggested by F. Kast and J. Rosenzweig which read that the contingency approach strives to perceive the interrelationships within and between subsystems as well as between the organization and its environment, and determine the types of relations or configurations of variables. It places emphasis on the multivariate nature of organizations and tries to make out how the organization operates in a changing environment and under specific circumstances. The contingency approach is used with a view to suggesting the designs of organization and management systems most suiting specific situations (Kast and Rosenzweig, 1972).

An analogous definition is given by F. Luthans and T. Stewart who believe that "the contingency approach deals with functional relationships between explicitly and operationally-defined system variables, i.e. situational, management and performance criteria variables" (Luthans and Stewart, 1977, 183).

Though the above definitions are not very pithy, we have selected them because they reflect both the common features and differences of the two approaches—system and contingency ones.

Indeed, the purpose of systems analysis is to identify the inter-relationships of all system variables and to present all interacting units as a system. The subject matter of systems analysis, from this point of view, is much wider than that of contingency analysis and fully embraces the latter.

While systems analysis is concerned with qualitative, basic relationships, exploring them in-depth, contingency analysis deals primarily with the quantitative side of relationships, and not all but the most essential ones (of course, from the researcher's point of view). From this, one may infer the similarities and differences of the approaches.

As a matter of fact, the interrelationships of elements, treated within the framework of the contingency approach (technology, external environment, organizational structure) were exposed and proved by systems analysis. A real or alleged novelty of the contingency approach is that it claims that it is possible to maintain a strict quantitative balance among the parameter values which ensures the maximum organizational effectiveness.

This statement is largely proved empirically through the study of actual relationships of different variables, their statistical processing and analysis. Fundamental empirical research has been conducted by T. Burns and G. Stalker (1961), A. Chandler (1962), J. Woodward (1965), J. Lorsch and P. Lawrence (1969).

The research showed that the firms that restructure their management systems taking due account of changes in technology and the environment are better off than those disregarding these factors. This research provided a basis for a number of hypotheses on the existence of more or less strict quantitative and qualitative relations among the factors of organizational development corroborated by other authors and management practice.

In the 1960s the USSR conducted a large-scale research project, on an analogous conceptual basis, with a view toward exploring quantitative relations between the technological and volume factors of production, environmental parameters and characteristics of the management system (Slesinger, 1976; Rapoport, 1968; Vorotnikova et al., 1970). Thousands of enterprises from different branches of the national economy were surveyed. The statistical data obtained convincingly showed a close relationship between the key technology factors, scale and type of organizations, and kind of organizational interaction with the external environment on the one hand, and a number of quantitative characteristics of management, on the other. We were even able to construct and practically test the normative models representing these relations (Development of staff standards and standard management structures, 1972).

One would think that the general logic and the impressive results of individual research support the sound prospects of the contingency approach and lay the basis for serious development of organization theory, especially for the use of quantitative criteria in substantiating and selecting organizational decisions. In spite of considerable research activities in the field, however, the expectations have not been realized so far.

First of all, the majority of the research, large though its scale and complex though the methods of statistical data processing may be, produce rather trite conclusions from the point of view of knowledge obtained on the basis of systems analysis. For example, the results of the extensive research conducted on the basis for a reliable frame of reference (Galbraith, 1973; Schoonhoven, 1981) led to the formulation of the following hypotheses:

1. The greater the technological uncertainty, the greater the positive impact of destandardization on effectiveness.
2. The greater the technological uncertainty, the greater the positive impact of decentralization on effectiveness.
3. The greater the technological uncertainty, the greater the positive impact of professionalization on effectiveness (Schoonhoven, 1981, 355-356).

It can be easily seen that the conclusions are rather universal, do not relate directly to specific environments, and may be proved on the basis of well-known logic from the systems approach.

But when the theoreticians of the contingency approach jump to more explicit and specific conclusions, their correctness is not subject to practical validation. The literature indicates that not a single empirical research study has been repeated and the results of the two studies coincided. Of course, there are always ready explanations for the emerging deviations (different objects under study, differences in the studied parameter definitions, etc.). But still, there has been no substantive advance so far regarding the hypotheses described above.

The critics of the contingency approach find it to be incorrect in one of the most important methodological aspects—inadequate account is taken of management influence on the controlled entity and environmental components (Longenecker and Pringle, 1978; Schreyogg, 1980). In fact, the majority of research explores only the direct impact of technology and the external environment on the necessary changes in organizational structure and organizational behavior. The question regarding the extent to which subjective managerial behavior can influence other internal and external elements or, at least, neutralize their impact has been studied inadequately.

There is no single answer to this question. If one considers quite concrete cases determining the organizational functioning at a certain instant in time, then the behavior of the system's active elements is undoubtedly a function of the state of its passive components and environment as relatively independent variables. Therefore, with regard to the fact that managerial relations constitute an important factor determining organizational behavior, studying the dependence of structural characteristics on the changes in the previously independent variables, making up the core of the contingency approach, is quite justified.

If one treats the situations in a broader context, as some processes of organizational development occurring during the time sufficient for management to realize its influence on the passive elements and environmental relations, then the notion of "independent variables" becomes unconstructive. This is confirmed by the latest trends in organizational development where an ever-important role is played by the flexible manufacturing systems which are highly sensitive to control action, and by adaptive management structures which are effective over a wide range of environmental changes.

Thus, depending on the problem, it is possible to take larger or smaller account of interrelationships between subjective and objective factors which determine the organization's behavior. Keeping in mind, however, that the final purpose of analysis is a search for ways to increase organizational effectiveness, it would be erroneous to omit from consideration the impact of management on all other system components.

It is possible in this case to determine quantitative and rigorous qualitative relationships among all considered variables of the organization in order to obtain criteria necessary for substantiating decisions and accurately forecasting changes in organizational effectiveness. The adepts of the contingency approach believe that this task may be solved, but the major problem is the uncertainty regarding "independent" variables. Thus far, no methods of problem solution in such conditions have been devised.

General contingency theory has arisen as a metatheory which "attempts to integrate the more specific or limited theories into a unified body of knowledge" (Luthans & Stewart, 1978, 684). It is also viewed as an "orienting strategy suggesting ways in which a phenomenon ought to be explained" (Schoonhoven, 1981, 350). It is expected that the respective disciplines and research will contribute to accumulation of concrete knowledge that could be systematized and ordered in order to acquire a qualitatively new and formalizable notion about the relationships and processes occurring within an organization.

The feasibility and expediency of developing a new metatheory in the form of general contingency theory are highly doubtful. It is generally recognized that the part of a metatheory is taken by the systems approach based on general systems theory which has proven itself as a methodological basis for a variety of applied systems research (Sadovsky, 1980). As for the specific knowledge that must be integrated within the framework of the suggested general contingency theory, it is much more concrete and is directly connected with the structure and behavior of a definite entity—organization. Therefore, we hold the view that it would be more correct to treat the contingency approach as a methodological element of organization theory. Organization theory integrates knowledge drawn from many special disciplines, and it can be shown that it has its own subject matter, notably, complex management relations (Rapoport, 1982).

5 METHODOLOGICAL BASIS FOR THE ANALYSIS OF MANAGEMENT RELATIONS

The behavior of each active element of a human activity system is determined by the totality of its relations with all other components and manifests itself in the form of relationships—actions providing for an exchange of material-energy resources and information required for their goal-oriented functioning (Rapoport, 1979).

The central role here is played by management relations which are behind the goals assigned to each element, the motives of their accomplishment, the scope of influence. All other relationships among the elements of a system, to the extent that they influence organizational behavior, are realized through management relations.

The key aspects are distinguished in management relations—organizational and economic. The former are stipulated by the role of each element in the process of decision making and implementation and regulated by the organizational structure. The economic relations characterize the pattern of resource utilization and the distribution of products and services. The economic relationships are determined by economic standards, prices, rates of employee remuneration, and material incentives.

The organizational and economic management relations are exercised by one and the same active element of the system, and the relationships realizing them often manifest themselves in one and the same actions—for example, in complex decision making. Therefore, though more often than not, the organizational and economic structures are established and changed independently, it makes sense to consider complex organizational-economic relations and the respective structures thereof.

Thus far, no description and analysis of the complex organizational-economic structure have been made; each structure was treated separately and their consistency was judged not on the whole but by individual elements. If, however, one considers the structure as a characteristic of an integral system determining the behavior and its effectiveness, then one may conclude that the isolated study and alteration of the two separate relations (organizational and economic) and relationships of individual elements are fruitless. Partial alterations, even if they seem justifiable, while solving some problems, generally engender others and do not enhance the accomplishment of the final result—higher organizational effectiveness.

Is it possible to give a sufficiently strict description of a complex structure in order to provide the basis for comparative interorganizational analysis? The inferences following from the systems approach to organizations make it possible to answer this question positively. We proceed from the following considerations.

First, given the ability of social systems to self-organize, we may assume that it is necessary and sufficient to describe only the invariant part of the structure. This makes it unnecessary to explore the content and sequence of the majority of concrete relationships stipulated by management relations.

Second, in describing these relations it suffices to identify only those factors that stipulate the behavior of the elements from the standpoint of their interaction influencing the general organizational effectiveness. This means that what should be subject to regulation is not the contents of activities but their results, and of those only that may bear upon the general results of system functioning and development.

Third, it is very important to define management relations by those characteristics that are not universal for all situations and may be directly related to the parameters characterizing the organizational technology and its relationships with the environment.

With due regard to the listed constraints, it is quite possible to describe the organizational and economic relationships and their interactions with basic characteristics such as responsibility and authority.

Even when these characteristics are used to describe organizational relations, they also, and to a considerable degree, determine the economic structure, since the distribution of roles in the process of decision making related to resource consumption and distribution of goods and services equally characterize both aspects of management relations. If we introduce into the definitions of responsibility and authority the parameters characterizing the economic side of organizational functioning, then they may well meet all the requirements.

The remarks suggest that the organizational-economic structure can be described in terms of responsibilities and authorities of organizational components in the form of "subject-object" matrices.

The responsibility matrix must include the following parameters:

- strict and complete definition of the goal, including constraints whose accomplishment is assigned to a concrete element (subsystem) of the organization;
- hierarchical level (officer) responsible for the quantitative formulation of the goal;
- definition of specific rules of behavior in case of contingencies and a procedure for the delegation of responsibility;
- determination of resources allocated for the goal accomplishment (area of administrative control, rates of resource utilization or profitability of their consumption);
- determination of specific administrative responsibility for a failure to achieve the goal, or for a deviation from the specifically assigned rules of behavior;
- determination of economic responsibility for the performance results (compensation of damages inflicted on other elements of the system in case of breaching the established constraints or violating the commitments, terms of utilization of extra-budgetary resources).

The authority matrix will contain the objects or activities with respect to which the element (officer) is authorized to make decisions:

- goals and objectives assigned to the reporting level and defined quantitatively;
- composition and volume of allocated resources, rates, and terms of their utilization;
- composition and volume of resources and operating results allocated for direct consumption by the active component (wages fund, profit share, other resources), terms of their consumption.

The responsibilities and authorities thus established stipulate the entire set of activities connected with implementation of organizational objectives and completely determine the part of the elements' behavior that is related to their goal-seeking interrelationships. An advantage of such a description of organizational structure is that it rather accurately characterizes both the organizational and economic management relations as regards their invariant part. Also, it does not require consideration of composition, content, and sequence of communications, nor the functions and procedures realizing them, which may be quite diverse and in many cases are determined by subjective factors.

The listed parameters of responsibility and authority can be easily modeled, and the correctness of their distribution is subject to formal analysis. For example, such an analysis is based on a balance of goals and resources distributed among the system elements regarding not only the descriptive but also the quantitative aspects. This balance provides a framework for partial organizational decision models following the setting of concrete goals at any level whose integration makes it possible to analyze the correctness and balance of authority distribution within the management system.

REFERENCES

1. Afanasyev, V. G., *Scientific Management of Society* (Systems case study), 2nd ed., Politizdat Publishers, Moscow, 1973.
2. *American Theories of Management* (Critique), ed. B. Z. Milner, and Ye. A. Chizhov, Mysl Publishers, Moscow, 1978.
3. Burkov, V. N., *Fundamentals of Mathematical Theory of Active Systems*, Nauka Publishers, Moscow, 1977.
4. Burns, T., Stalker, G., *The Management of Innovation*, Tavistock, London, 1961.
5. Chandler, A., *Strategy and Structure*, MIT Press, Cambridge, MA, 1963.
6. *Development of Staff Standards and Standard Management Structures for Enterprises and Production Associations. Guidelines.* Research Institute of Labor, Moscow, 1972.
7. Drucker, P., *Management: Tasks, Responsibilities, Practices.* John Wiley & Sons, New York, 1974.
8. Galbraith, J., *Designing Complex Organizations*, Addison-Wesley, Reading, MA, 1973.
9. Gvishiani, J. M., *Organization and Management*, 2nd ed., Nauka Publishers, Moscow, 1972.
10. Haynes, W. M., and Massie, J. L., *Management*, 2nd ed., Prentice-Hall, Englewood Cliffs, N.J., 1969.
11. Kast, F., and Rosenzweig, J., General Systems Theory: Application for Organization and Management, *Academy of Management Journal*, December, 1972, pp. 453-463.
12. Kast, F., and Rosenzweig, J., *Organization and Management: A Systems Approach*, 2nd ed., McGraw-Hill, New York, 1974.
13. Kuhn, A., and Beam, R., *The Logic of Organization*, Jossey-Bass Publishers, San Francisco, 1982.
14. Lawrence, P. R., and Lorsch, J. W., *Organization and Environment.* Richard D. Irwin, Homewood, 1969.
15. Litterer, J. A., *The Analysis of Organizations*, John Wiley and Sons, New York, 1972.
16. Longenecker, J. G., and Pringle, C. D., The Illusion of Contingency Theory as a General Theory, *Academy of Management Review*, v. 3, No. 3, 1978, pp. 679-682.
17. Luthans, F., and Stewart, T. I., A General Contingency Theory of Management, *Academy of Management Review*, Vol. 2, pp. 181-195, 1977.
18. Luthans, F., and Stewart, T. I., The Reality or Illusion of a General Contingency Theory of Management: A Response to the Longenecker and Pringle Critique, *Academy of Management Review*, 3/3, pp. 683-687, 1978.
19. Meyer, M. W., and Associates, *Environments and Organizations*, Jossey-Bass, San Francisco, 1978.
20. Milner, B. Z., Evenko, L. I., Rapoport, V. S., *Systems Approach to Management Organization*, Ekonomika Publishers, Moscow, 1983.
21. Miner, J. B., *Theories of Organizational Structure and Process*, Macmillan, New York, 1982.
22. Ovsiyevich, B. L., *Models of Organization Structure Formation*, Nauka Publishers, Leningrad, 1979.
23. Rapoport, V. S., *Managerial Performance Standards at Industrial Enterprises*, Nauka Publishers, Moscow, 1968.

24. Rapoport, V. S., *Innovation Management in Industry*, Ekonomika Publishers, Moscow, 1979.

25. Rapoport, V. S., Methodological Problems of Systems Research into Organizational and Economic Aspects of Management, *Systems Research. Methodological Problems.* Yearbook, Nauka Publishers, Moscow, pp. 159-181, 1982.

26. Sadovsky, V. N., Systems Approach and a General Systems Theory: Status, Basic Problems and Prospects of Development, *Systems Research: Methodological Problems.* Yearbook, Nauka Publishers, Moscow, pp. 29-54, 1980.

27. Schoonhoven, C. B., Problems with Contingency Theory: Testing Assumptions Hidden within the Language of Contingency "Theory", *Administrative Science Quarterly*, 26, pp. 349-377, 1981.

28. Schreyogg, G., Contingency and Choice in Organization Theory, *Organization Studies*, 1/4, pp. 305-326, 1980.

29. Slesinger, G. A., *Labor in Managing Industrial Production*, Ekonomoka Publishers, Moscow, 1967.

30. Vorotnikova, V. V., Pavlenko, A. P., Slesinger, G. A., *Performance Standards for Engineering and Office Staff*, Ekonomika Publishers, Moscow, 1970.

31. Woodward, J., *Industrial Organizations: Theory and Practice*, Oxford University Press, London, 1965.

DISCUSSION OF RAPOPORT'S PRESENTATION

Umpleby

What is the motivation that underlies your work?

Rapoport

I assume that management in social systems is a universal subject for systems analysis. It covers all problems from relations between states to relations between firms, their divisions and individuals. I am trying to identify the key factors influencing the behavior of people and groups in social systems. My problem is: how to manage organizational behavior by regulating the relationships among participants in the decision-making process? I think that in different social systems (the U.S. and the Soviet Union) there are different means of regulating relationships but all of them are based on similar principles and rules.

Umpleby

In the U.S. we have been moving away from modeling as a method of problem solving to using group process methods to "orchestrate" a conversation among a group of people who themselves define a set of problems and create solutions.

Rapoport

We read the systems literature and use the systems approach. But these techniques have limits. They are expensive. All decisions cannot be taken this way. Other decisions require a more formal approach. It would be unrealistic for us to use these all of the time.

When there is a formal structure, there is a need for a less formal component.

Mitroff

In my consulting practice I find that managers are understanding more all the time that they cannot organize in a bureaucratic way. The structures do not respond rapidly enough. If they want to be more competitive, they realize they have to change toward a more informal structure of small teams. This is a strong trend in the U.S. and Europe.

Rapoport

The current system has a lot of inertia. These are the most urgent issues now, and they are being discussed at the highest levels in the USSR. To maximize effectiveness an organization must have the most highly bureaucratic structure. But if the environment changes, the structure must change. If we optimize the present structure, we make change difficult. Every manager resolves this difficulty on an *ad hoc* basis. There are about half a million managers in the highest categories in the USSR. The top people are not necessarily the best analysts. Management is a mix of science and art. There is a need for methods to help managers. Scientific rigor must yield to practical considerations. Managers must be convinced that they need the new approaches.

CHAPTER

12

MANAGING COMPLEXITY IN THE SYSTEMS AGE: A RENAISSANCE SYSTEMS PERSPECTIVE

William J. Reckmeyer

1 INTRODUCTION

Most of us are probably well aware of the recent emphasis here in the West on business and the often fervent search for excellence that now extends beyond the corporate world into virtually every aspect of modern life. Disturbed by the increasing volatility of societal affairs, enamored with emerging high technologies, and spurred on by wave after wave of best-sellers by academic specialists and management evangelists alike, people here and abroad have been taking a long look at organizational life and making changes that may signal a profound transformation in the way we work. This renewed commitment to quality is not unprecedented, of course, for there has been a disciplined interest in improving organizational performance at least since the turn of this century. Nevertheless, it does reflect a growing realization that times are changing and that "business as usual" won't work as well in the future as it has in the past. The barriers to excellence are legion, though, and overcoming them with an appropriate combination of character and competence requires that we begin developing means that are better suited to the realities of modern times.

Even the best people are hard-pressed to figure out what to do, however, given the turbulence of our world and the general absence of proven guidelines for succeeding under such conditions. Unfortunately, most of the management advice being hyped today provides more inspiration than it does solid help. The basic message is fairly straight-forward: learn from the winners and emulate their techniques if we want to duplicate their success (eg., Bennis and Nanus, 1985; Brown and Weiner, 1984; Drucker, 1980; Hickman and Silva, 1984; Kanter, 1983; Peters and Waterman, 1982; Pinchot, 1985; Toffler, 1985). What works for one organization rarely works exactly the same for others, despite the universal appeal of anecdotes and platitudes, for everyone's situations and abilities are different. I haven't discovered any magic answers, but have found that we need more integrative approaches that are longer on both theory and praxis—that is, pragmatic philosophies which enhance our individual and collective ability to handle complexity more effectively and responsibly. Considerable progress has been made in our field during the last several decades towards developing such approaches and it seems apropos that we briefly clarity its relevance to these issues at this initial meeting of the US-USSR Project on the Foundations of Cybernetics and General Systems Theory.

2 THE SYSTEMS AGE

Scholars and pundits have been noting for some time now that humanity seems to be in the dawn of a new age that is quite different from the life experienced by previous generations. People have different names for it depending on which features interest them the most—some call it the Post-Industrial Society, the Nuclear Age, or the Information Age; for others, it is the Solar Age, the Third Wave, or the Age of Aquarius—but all argue rather convincingly that we are at a major turning point in human history (eg., Bell, 1973; Boulding, 1964; Capra, 1982; Ferguson, 1980; Fuller, 1981; Henderson, 1981; Toffler, 1980). The focus on a few key factors is understandable, but the most momentous aspect of modern society is the multiple systems of factors that are synergistically producing an increasingly complex Systems Age so marked by turmoil and death-bets that we find it tougher and tougher to keep our heads above water, much less to anticipate and deal with whatever might be coming around the next bend (eg., Ackoff, 1981; Beer, 1975; de Rosnay, 1979; Reckmeyer, 1982). Changes of this sort are not unparalleled in history, as anyone familiar with the Enlightenment and the Industrial Revolution would be quick to point out. Nor are they as sweeping as many modern prophets seem to think, for most of the world remains mired in the Agricultural Age and only its more advanced regions have actually begun making the transition from the Machine Age into the Systems Age. Still, it is clearly a time of extensive transformation and thus full of opportunities as well as dangers. The challenge facing us all, therefore, is how best to maximize the former while minimizing the latter.

Successful management has always necessitated a subtle blend of visionaries and missionaries—people who know where they want to go and how to organize their efforts to get there—and has never been as easy as some experts would lead us to believe, even in the simplest or stablest of situations. Management in the Systems Age poses even greater difficulty, for contemporary issues are far more complex than the ones we are used to handling. The issues confronting us today seldom respect the neat disciplinary boxes we learn in school or the functional domains we encounter at work. They are not reasonably well-defined problems that can be solved once and for all, as much as they are ill-defined messes that must be managed over long periods of time. This is especially disconcerting since many of the solutions we devise are often worse than the original conditions we seek to redress. Our dilemma is that we know more about solving problems than we do about managing messes. As such, we tend to treat every issue with the overly mechanistic approaches that have worked so well is the past, even though they may be inappropriate for the situation at hand—much as the drunk does when looking for his keys under the light.

Developing the character and competence to succeed on an individual basis under such conditions is no easy matter, to say the least, but it is even more difficult to do on a collective basis. There are obviously notable exceptions, but most organizations tend to cope with changing conditions in an ad-hoc fashion—fighting fires as needed rather than developing better means for anticipating and preventing them in the first place. Moreover, they generally devote most of their attention and resources to those matters that are directly related to specific products in a never-ending pursuit of greater profits and returns on investment, but spend comparatively little on those less obvious matters that are essential to their own organizational viability. There are no recipes for excellence, even from systems people; but we do know that the healthier people are, the greater is their generic ability to handle turbulent conditions. The same is true for organizations, although we know far less about organizational health than we do personal health and even less about what we can do to achieve and maintain it.

The principal obstacles to healthy organizations, it seems to me, derive from the piece-meal fashion in which most of us deal with the world and our general inability to work together as well as current conditions require. My research and experience, reinforced by that of others, demonstrate that real innovation is the result of a creative tension between convergent and divergent modes of thought and action. We have to open up in order to expand our horizons, but cannot open up too much or we find it difficult to get much done. The Law of Requisite Variety reminds us that only variety can handle variety, which means that we must either dampen the complexity of the situation at hand or amplify our own. We tend to do the former more than the latter, but messes are harder to dampen than problems and we need approaches that amplify our own abilities rather than vice versa. The Law of Requisite Parsimony reminds us that too much diversity hinders more than it helps, so we must develop approaches that help us focus as well as amplify our efforts. The Law of Requisite Saliency reminds us that some aspects of a situation are more important than others, so we need approaches that help us separate the wheat from the chaff. The Law of Requisite Flexibility reminds us that everything cannot be done at once, so we need approaches which maximize freedom and help us zero in on our goals in an increasingly refined fashion (Ashby, 1956; Reckmeyer, 1984; Warfield, 1986).

Machine Age approaches to complex matters grew out of the fascination of 17th-century scholars with simple machines and the enormous success of methods that were initially developed in the 18th-century physical sciences. Characterized by a strong belief in the ultimate objectivity of knowledge,an emphasis on isolated and static views of phenomena, and a commitment to systematic means of investigation, these approaches have generally stressed the differences among phenomena and led to an increasing reliance on specialization as the dominant form of thought and action here in the West. Systems Age approaches grew out of the belief by a growing number of 20th-century scholars and practitioners that more traditional approaches are inadequate for treating complex biological and social issues even though they have been extremely powerful for addressing simpler physical and mechanical concerns. Characterized by a profound belief in the intrinsic subjectivity of knowledge, an emphasis on more contextual and evolutionary views of phenomena, and a commitment to a blend of rational and intuitive procedures, these approaches have usually stressed the similarities among phenomena and encouraged the need for greater generalization so that people can better integrate their thoughts and actions.

Nevertheless, many Systems Age approaches still over-emphasize the development of specific tools and techniques for managing complexity rather than the sorts of pragmatic philosophies that I think are necessary for success. This is not too surprising, given our natural concern for the here-and-now at the expense of the there-and-then, but is no less disturbing in light of how significantly conditions are changing. Our primary concern, after all, remains pretty much the same as it has since the dawn of our species: doing everything we can to better understand and deal with the "buzzing, blooming confusion" of the world we experience. Humanity has made considerable progress toward this end during its relatively short tenure here on Earth, largely because Homo sapiens is a race of superb generalists. We are going to have to improve by leaps and bounds in the years to come, though, if our success is going to continue as it has in recent centuries. In sum, we need dreamers who know how to make dreams come true in a collective setting and more integrative approaches with a better combination of rigor and imagination to help us amplify and focus our efforts to do this in meaningful ways.

3 A RENAISSANCE SYSTEMS PERSPECTIVE

In their most elementary sense, approaches are the ways in which people handle reality. For most of us, these approaches are highly idiosyncratic and largely implicit: we cope as best we can and rarely give much thought to how we do it. In a more rigorous sense, however, approaches are seen as much more shared and explicit: they are paradigms that help us perceive, understand, and deal with our world in more structured fashions (Kuhn, 1970). As such, I have found it important to realize that there are three critical features to all paradigms: epistemological ones regarding how we are able to know the world we inhabit; ontological ones regarding what we know about that world; and methodological ones regarding how we interact with it—recognizing that we ourselves are an intrinsic part of that world (Reckmeyer, 1982). The epistemological features concern the nature of observers and help us clarify our role in the generation of knowledge. The ontological traits concern the nature of observed phenomena and help us explain what we see in the context of the universe as a whole. The methodological features concern the nature of observations and help us develop blueprints for action. The extent to which they are consciously manifested can vary considerably from one approach to another, of course, but all three are characteristic of any paradigm and should be explicitly integrated as much as possible.

Reflecting humanity's innate inability to make sense of everything all at once, scholars and practitioners have developed various paradigms that essentially approach the world in terms of parts and wholes. Machine Age paradigms have usually treated wholes as equal to the nature of their parts and assumed the existence of an objective reality that can be fully known by human observers once we determine the basic laws of Nature. Epistemologically reductionistic, ontologically mechanistic, and methodologically analytic, these paradigms have proven extremely valuable for addressing the harder kinds of issues (harder in the sense that they can be satisfactorily treated with great precision, not in the sense of being more difficult to address). Systems Age paradigms usually treat wholes as different than their parts and propose a subjective world that can never be completely known by any observers because the fundamental laws of Nature are always contextual. Epistemologically expansionistic, ontologically holistic, and methodologically synthetic, these approaches have been developed to deal with the softer kinds of issues that have proven less amenable to more conventional approaches (softer in the sense that they cannot be satisfactorily addressed with a high degree of precision).

Respective critics and proponents of both kinds of approaches have been arguing the relative merits of their own views for some time now, usually in absolutist terms that promote their candidate paradigms as the only worthwhile means for addressing the world. Most systems people, for example, characterize science as a paradigm for simplicity and systems science as a paradigm for complexity. Life is not so easily dichotomized, however, for people have to wrestle with a range of hard-medium-soft conditions across a range of meta-macro-micro levels and we benefit most from those approaches that encompass a comparable spectrum of multi-dimensional capabilities. Rather than simply characterizing our field as an exclusively holistic antithesis to the predominantly mechanistic thesis of rational western thought, it is far more suitable to view systems science as a comprehensive meta-paradigm that subsumes the harder traditional approaches as well as the emerging softer ones—much as Newtonian mechanics is now seen as a special case of Einsteinian physics. Although the systems community as a whole has not really recognized this state of affairs, much less woven the sort of coherent synthesis that I believe is possible at this time, I think we can discern a fairly substantial skeleton amidst the flesh created during the first forty years or so of the field's existence.

Driven by this realization and the needs of my own work as a scholar, educator, and practitioner, I have been developing such a synthesis—what I have been calling renaissance systems science or renaissance cybernetics, to highlight its explicit comprehensive and integrative nature—and have been using it with various public, private, academic, and governmental clients here in the United States and Europe during the last fifteen years. It draws upon and extends the generic theoretical insights and practical skills of systems science (a general art/science of organized relationships) and cybernetics (a general art/science for regulating relationships); more specific knowledge and tools from related sub-fields (systems dynamics, systems management, systems engineering, control engineering, informatics, operations research, game theory, etc.); and highly specialized contributions from the more traditional disciplines and professions (Reckmeyer, 1982 and 1984). As such, it deliberately encompasses and manifests a unified epistemology, ontology, and methodology commensurate with the entire panoply of human affairs and helps people become exactly the sorts of pragmatic philosophers and reflective practitioners society requires as it moves into the Systems Age.

Epistemologically, renaissance systems science recognizes the inherent subjectivity of all observers and stresses that knowledge is constructed not received—the result of an inquisitive rather than an acquisitive process. Rejecting what I call the Dogma of the Immaculate Perception, cyberneticians and a small number of biologists and cognitive scientists are beginning to emphasize what some philosophers and social scientists have argued for years: that knowledge consists of myriad distinctions drawn by observers, not absolutes given by God (eg., Bateson, 1972; Berger and Luckman, 1967; Boulding, 1956; Checkland, 1981; Clemson, 1984; Maturana and Varela, 1980; Pask, 1976; Piaget, 1959; Reckmeyer, 1982; von Foerster, 1982; Watzlawick, 1984). Our knowledge clearly involves a range of hard-medium-soft and meta-macro-micro distinctions; but in the deepest sense, the universe is made up of stories not atoms. Perception is reality. All enquiry, whether formal or informal, is an inherently purposeful activity biased by our previous experience and underlying attitudes, values, and beliefs—so we should always be as deliberately self-referential as possible in everything we do. In short, the field helps us be more comfortable with uncertainty because the process of observation necessarily limits what can be known about observed reality—which is the very point of the General Uncertainty Principle (Reckmeyer, 1982).

Ontologically, the field recognizes the essential interdependence of observed phenomena and emphasizes the relational patterns that connect rather than the substantive differences that separate. It posits that all aspects of reality (physical, biological, social, technological, noological) can be better understood when treated as systems: wholes composed of parts that exist in larger environments and display a diversity of emergent characteristics that derive from the way they are internally/externally organized and change through time (eg., Ashby, 1956; Bateson, 1979; Beer, 1977 and 1980; Bogdanov, 1980; Boulding, 1978; Bowler, 1981; Checkland, 1981; Cook, 1980; Corning, 1983; de Rosnay, 1979; Forrester, 1968; Iberall, 1972; Jantsch, 1980; Kuhn, 1974; Laszlo, 1972; Lovelock, 1979; Maturana and Varela, 1980; Miller, 1978; Odum, 1983; Prigogine and Stengers, 1984; Rapoport, 1986; Reckmeyer, 1982; Sheldrake, 1981; von Bertalannfy, 1968; Weinberg, 1975). There is clearly a range of hard-medium-soft and meta-macro-micro phenomena, but the field holds that all systems are the result of certain common organizing processes and thus manifest similar traits and behaviors. It pays special attention to the variety of linear, mutual, and circular causal relationships constituting phenomena and the various regulatory processes that enable some of them to pursue sundry purposeful activities. In short, it helps us be more comfortable with incompleteness because an observed world of infinite context can never be fully understood through a process of finite enquiry plagued by rampant uncertainty—which is the very point of the General Incompleteness Principle (Reckmeyer, 1982).

Methodologically, the systems paradigm recognizes that observational blueprints are essential to sophisticated enquiry and clarifies some broad guidelines and operational procedures for meaningful action that enable people to move from general matters to more specific ones in a disciplined way. There is clearly a range of hard-medium-soft and meta-macro-micro tools and techniques for doing this, regardless of whether the action involves research or intervention, but it requires a blending of theory and data (through a combination of abductive, deductive, and inductive efforts) that is consistent with the epistemological and ontological parts of the paradigm. It also provides systematic steps that detail how people can proceed in an organized and timely fashion. Finally, it includes a set of interactive group methods for implementation in a collective setting since people rarely work in isolation (eg., Ackoff, 1962 and 1978; Argyris, 1985; Athey, 1982; Beer, 1966 and 1985; Checkland, 1981, Churchman, 1971 and 1979; Clemson, 1984; de Bono, 1985; Feyerabend, 1975; Hardin, 1985; Kertzner, 1984; Klir, 1985; Linstone, 1984; Majone and Quade, 1980; Miser and Quade, 1985; Morgan, 1983; Olsen, 1982; Reckmeyer, 1982; Sage, 1977; Schon, 1983; van Gigch, 1978; Vickers, 1965; von Oech, 1982 & 1986; Waddington, 1977; Weinberg, 1985). In short, it helps us be more comfortable with complementarity because the uncertainty and incompleteness of enquiry preclude excessive reliance on singular means for doing things—the very point of the General Complementarity Principle (Reckmeyer, 1982).

Our field is young and extremely broad, however, and it would be misleading to imply that there is unanimity concerning even its basic scope and contributions—much less agreement on the sort of synthesis I have sketched out here. Some people are chiefly scholars, while others are mainly practitioners. Some focus on epistemological issues, some on ontological matters, some on methodological ones, and some on specific areas of research and application. Some pursue very broad philosophical concerns and ignore the messy details, while others zero in on particular ideas or tools and never bother with the big picture. There is even disagreement, often strident, on whether we should call ourselves systems scientists or cyberneticians. Thus, the picture one creates of this elephant, whether we are inside or outside the field, changes substantially depending on where we sit and who we talk with (eg., Blauberg, 1977; Buckley, 1968; Gray and Rizzo, 1973; Klir, 1972 and 1978; Krippendorf, 1979; Lasker, 1981; Lilienfeld, 1978; Machlup and Mansfield, 1983; Meadows, 1982; Miser and Quade, 1985; Reckmeyer, 1982; Richardson, 1984; Trappl, 1983 and 1986; von Foerster, 1974). The key is to remember that the field is a whole elephant, even if we cannot see it all at one time, and not merely a collage of trunks and tusks and tails.

In essence, I treat systems science as a sophisticated Rosetta Stone—a meta-paradigm with meta-epistemological, meta-ontological, and meta-methodological features that are relevant for all observers, all observeds, and all observations—which is the very point of the General Relativity Principle (Reckmeyer, 1982; von Foerster, 1982). Academically, the field is principally concerned with integrating human knowledge to enhance our understanding of the universe. Professionally, it is primarily concerned with using this knowledge to improve that world. The advantages of a more unified approach can be enormous, chiefly in terms of its relevance to the full range of issues that interest and affect humanity. Resolving these issues requires expertise from every domain of human experience, but is not easy to pull together given the myopia and intransigence that characterizes most disciplines and professions as well as the shear volume of highly specialized knowledge the species has generated during its existence. We need more integrative paradigms like renaissance cybernetics that encourage cross-fertilization across different fields while minimizing the attendant dangers of information overload. Moreover, we must realize that the main value of systems science lies in its (potential) nature as a unified whole and not in its specific insights and skills alone—no matter how valuable they may be in specific situations.

4 MANAGING COMPLEXITY

From a systemic perspective, the most significant difficulties facing modern organizations involve what they can do to manage their changing relationships in ways that will ensure both their short-term and long-term viability without seriously compromising the viability of their members or the environments in which they exist. This is not too challenging when conditions remain fairly stable or even when changes are reasonably gradual; people have ample time to figure out what is happening and eventually develop an appreciation of what it takes to succeed. When conditions begin changing so fast that people have no real sense of what is going on and the consequences for failure escalate, however, they are far less able to sort out what needs to be done. As a result, we typically place a premium on maintaining the status quo and reactively deal with new situations in an ad-hoc, piece-meal manner that over-emphasizes basic operational matters at the expense of less tangible strategic ones. Even more importantly, we often lose sight of the ethical dimensions of what we are doing and focus our attention exclusively on being as effective and efficient as we possibly can—behavior that is ultimately unhealthy for everyone involved when carried to an extreme.

These attitudes and behavior make it rather problematic for most organizations to develop the foresight and flexibility necessary for excellence in turbulent times and severely limit the extent to which they can accommodate change, much less turn it to their advantage. Organizations have always had to manage two kinds of matters: substantive ones, which are their principal reasons for being; and organizational ones, which affect their abilities to properly handle these substantive concerns. Success in the Systems Age still requires progress in both of these provinces, but now it means that organizations must place a much greater emphasis on anticipating changing conditions and proactively balancing concerns for the here-and-now with the there-and-then. In systems terms, this means they must manage two kinds of change: morphostatic changes, which tend to maintain existing conditions; and morphogenic changes, which tend to transform them. The temptation is to develop specific approaches to specific issues and leave it at that, but our paramount goal should be to develop healthy organizations with generic abilities to manage specific issues as desired. This also means that organizations have to be accomplished learners as well as doers—explicitly able to develop relevant insights and skills that are appropriate to changing conditions through a combination of boot-strapping and reliance on approaches that have proven successful in the past.

Addressing any issue (hard-medium-soft, meta-macro-micro, problem-mess, etc.) requires the on-going performance of certain critical tasks. There are many systems blueprints available for doing this (often including marked differences in terminology, arrangement, detail, and emphasis), but all ultimately involve variations of several functions that are essential to any cybernetic activity: detection, selection, effection, connection, and correction (eg., Athey, 1982; Checkland, 1981; Clemson, 1984; Kertzner, 1984; Klir, 1985; Kuhn, 1974; Miser and Quade, 1985; van Gigch, 1978; Warfield, 1982). First, we have to detect what is going on: we have to formulate the situation, clarifying its principal features and our own purposes and biases in the enquiry. Second, we have to select what needs to be done: we have to develop some sort of plan to improve the situation, designing preferred alternatives and choosing a course of action that we think will work. Third, we have to effect the plan: we have to take some sort of action, doing whatever it takes to get the job done. Finally, we have to do this in a connected fashion: we have to iteratively monitor and adjust activities as needed—continuously detecting, selecting, effecting, and correcting as conditions change until we either achieve success or acknowledge failure. There are usually numerous sub-tasks within these basic tasks that enable people to implement the process at a desired focus and depth, but these are the minimum ones necessary for viable systems practice (Reckmeyer, 1984).

Any individual can use a systems approach to his or her great benefit, I think, but organizations have evolved into important societal institutions precisely because people have discovered the decided advantages that accrue to them from organized collective activity. We know quite a bit as a species about how to co-exist (despite the tensions between our two countries and the situation in places like Northern Ireland and the Middle East), but we do not really know as much about what it takes to work well together—even when we are operating in the same general context and with similar purposes. The heart of renaissance systems management, therefore, lies in its heavy emphasis on interactive uses of the approach by transdisciplinary teams of critical decision-makers, stake-holders, and experts to construct shared domains for thought and action. Most collective work occurs in meetings and most of these are an enormous waste of people's time and talent. Largely venues for either dispensing information or rubber-stamping decisions that have already been made by others, meetings are hardly ever as useful as they can and need to be because people know relatively little about groups and recent developments for enhancing group productivity.

Several attributes distinguish good interactive systems practice. One is its use of an integrated set of hard-medium-soft group methods, which enable people to move through the full process of management tasks in ways that both amplify and focus their efforts. Another is its reliance on facilitators skilled in the use of the systems approach as a whole (and especially these group methods) who help structure people's interactions so that they are free to concentrate on the work at hand. Other features include the use of operations rooms that are ergonomically designed to support such work on a regular basis and special computer capabilities for speedier record-keeping of the group's work as well as easier handling of numerous variables (eg., Doyle and Straus, 1976; Fisher and Ury, 1983; Goodman, 1986; Johnson and Johnson, 1975; Kertzner, 1984; Maier, 1977; Olsen, 1982; Reckmeyer, 1984; Robinson, 1984; Swap, 1984; Warfield, 1982 and 1986).

Five important benefits derive from such systems practice when it is done well. First, it helps groups make significant progress towards resolving issues that concern them through a more structured use of their time and effort. Second, it helps people continuously document what they are doing, which improves their operation in the moment as well as their development of a collective memory. Third, it helps them improve their understanding of the situation through their exposure to multiple perspectives and thus enhances their ability to implement their plans. Fourth, it helps people develop a better understanding of themselves and their colleagues, thereby improving their overall teamwork. Fifth, and most important of all, it helps groups buy into the decisions being made—for the best plans are those that are understood and supported by the people who are affected by and who have to implement them, not those delivered by God or the boss or even a knowledgeable systems expert (Reckmeyer, 1984; Warfield, 1982).

Now, the best definition I ever heard for management is the same one I use for life in general—"One damn thing after another"—and keeping on top of it all in today's complex organizations without using this approach in an interactive fashion makes about as much sense as playing football without huddling. The basic operating premise among systems people as well as most management experts is that planning is an essential part of management and that good planning is a prerequisite for proactive behavior (eg., Ackoff, 1981 and 1984; Argyris, 1983; Beer, 1977 and 1980; Checkland, 1981; Churchman, 1971; Clemson, 1984; DeGreene, 1982; Jantsch, 1975; Kast and Rosenzweig, 1985; Kertzner, 1979; Linstone, 1984; Morgan, 1986; van Gigch, 1987). Over-planning is not the answer, though, any more than crisis management is. Managing complexity generally necessitates a subtle blend of design and evolution, for planning is best seen as a domain for disciplined improvisation. Too much design inhibits the natural evolution of organizational concerns and limits flexibility, while too little design makes life a crap-shoot and limits organizational ability to pursue collective goals in a coherent manner. Moreover, the sorts of issues being addressed require gardeners more than they do mechanics. Problems can be fixed, but messes need nurturing and we should pay better attention to developing more systemic modes for intervention. We have to work with the natural order rather than always seeking to impose our will on matters for our own convenience (which is not only irresponsible, but may also be ineffective in the long run). No matter how much we amplify our own abilities, we can never manifest the requisite variety to control everything. So we should treat organizational issues and life in general with a combination of chutzpah and humility, regardless of how much power we might have to do as we please at any particular moment.

5 CONCLUSION

No approach is a panacea, as we all know, and there are no recipes for ensuring excellence no matter how simple the situation or how accomplished the people. Committing cybernetics in public is not always as easy as it might seem from these comments, either, for there is at least as much art as there is science in renaissance systems work. Nevertheless, humanity deserves and requires more unified approaches as it enters the Systems Age and one of the principal goals of this project as well as the field as a whole should be to spend more effort on developing the kind of synthesis I am advocating here. Space has precluded all but a brief review of its features, but a few points are worth emphasizing. First, it is a tightly integrated meta-paradigm whose systemic epistemological, ontological, and methodological features help people develop more appropriate means for perceiving, understanding, and tackling whatever concerns them in an organized fashion. Second, it explicitly includes a range of hard-medium-soft theoretical insights and practical skills that help us address the full range of human experience. Third, it consciously includes a full set of interactive group methods that help people manage complexity in a collective way. Fourth, its blend of generic and specific features help people move among the meta-macro-micro dimensions of a situation without getting too lost in the clouds or too stuck in the mud. Fifth, its comprehensive and integrative nature both encourages and facilitates better cross-fertilization among the various realms of human endeavor. Finally, it explicitly recognizes that it takes more than technical proficiency in the approach to ensure responsible results and thus calls for systemic applications of the paradigm—in the form of disciplined conversations, if you will—that improve people's ability to manage their own affairs, thereby developing autonomy rather than the kind of dependence which normally occurs when we rely on others to provide the answers for us.

In closing, I would also like to emphasize that it is generally a long way from rhetoric to reality and people need more than platitudes and pep talks—even from the systems field—to construct the kind of individual and collective excellence we desire. Judging by the numerous born-again organizations that have embraced the formulas for success being offered by the latest management experts, most of which treat symptoms rather than causes, it seems evident that band-aid management isn't the answer. Managing complexity ultimately involves wrestling with the substantive and organizational matters specific to whatever milieu people are working in and the mechanistic approaches that have worked well for solving problems rarely work as well for managing the messes that are increasingly characteristic of life in the Systems Age. Using the theoretical insights and practical skills of the systems field helps people learn how to better integrate their thoughts and actions, thereby developing the coherence necessary to ameliorate the oscillatory behaviors that are usually pathological for complex systems. Organizations will always need people who know relevant details, for sure, but the really vital need right now is for more reflective practitioners possessing the generic capabilities to sail in uncharted waters and not for more technicians armed with the latest recipes for excellence. In short, we need more people who are well versed in pragmatic philosophies like renaissance systems science and know how to use them to manage complexity with a better combination of character and competence.

SELECTED BIBLIOGRAPHY

Ackoff, Russell L. *The Art of Problem Solving: Accompanied by Ackoff s Fables.* New York: Wiley, 1978.

______. *Creating the Corporate Future: Plan or Be Planned For.* New York: Wiley, 1981.

______ et al. *A Guide to Controlling Your Corporation's Future.* New York: Wiley, 1984.

______ et al. *Scientific Method: Optimizing Research Decisions.* New York: Wiley, 1962.

Amara, Roy and Lipinski, Andrew J. *Business Planning for an Uncertain Future: Scenarios and Strategies.* New York: Pergamon, 1983.

Apostel, Leo et al., eds. *Interdisciplinarity: Problems of Teaching and Research in Universities.* Paris: Organization for Economic Co-operation and Development, 1972.

Argyris, Chris. *Reasoning, Learning, and Action: Individual and Organizational.* San Francisco: Jossey-Bass, 1983.

______ et al. *Action Science: Concepts, Methods, and Skills for Research and Intervention.* San Francisco: Jossey-Bass, 1985.

Ashby, W. Ross. *Design for a Brain.* New York: Wiley, 1960.

______. *An Introduction to Cybernetics.* New York: Wiley, 1956.

______. *Mechanisms of Intelligence.* Seaside: Intersystems, 1981.

Athey, Thomas H. *Systematic Systems Approach: An Integrated Method of Solving Systems Problems.* Englewood Cliffs: Prentice-Hall, 1982.

Bateson, Gregory. *Mind and Nature: A Necessary Unity.* New York: Dutton, 1979.

______. *Steps to an Ecology of Mind.* New York: Ballantine, 1972.

Bateson, Mary C. *Our Own Metaphor.* New York: Knopf, 1972.

Bayraktiar, B. A. et al., eds. *Education in Systems Science.* New York: Halstead, 1979.

Beer, Stafford. *Brain of the Firm.* New York: Wiley, 1977.

______. *Decision and Control: The Meaning of Operational Research and Management Cybernetics.* New York: Wiley, 1966.

______. *Diagnosing the System for Organizations.* New York: Wiley, 1985.

______. *The Heart of Enterprise.* New York: Wiley, 1980.

______. *Platform for Change.* New York: Wiley, 1975.

Bell, Daniel. *The Coming of the Post-Industrial Society.* New York: Basic, 1973.

Bennis, Warren and Nanus, Burt. *Leaders: The Strategies for Taking Charge.* New York: Harper & Row, 1985.

Berger, Peter L. and Luckman, Thomas. *The Social Construction of Reality: A Treatise in the Sociology of Knowledge.* Garden City: Doubleday, 1967.

Blake, Robert R. and Mouton, Jane S. *Managerial Grid III.* Houston: Gulf, 1985.

Blauberg, Ivan V. et al. *Systems Theory: Philosophical and Methodlogical Problems.* Moscow: Progress, 1977.

Bogdanov, Alexander A. *Essays in Tektology.* Seaside: Intersystems, 1980.

Boulding, Kenneth E. *Ecodynamics: A New Theory of Societal Evolution.* Beverly Hills: Sage, 1978.

______. *The Image: Knowledge in Life and Society.* Ann Arbor: University of Michigan, 1956.

______. *The Meaning of the Twentieth Century: The Great Transition.* New York: Harper & Row, 1964.

______. *The World as a Total System.* Beverly Hills: Sage, 1985.

Bowler, T. Downing. *General Systems Thinking: Its Scope and Applicability.* New York: North Holland, 1981.

Briggs, John P. and Peat, F. David. *Looking Glass Universe: The Emerging Science of Wholeness.* New York: Simon & Schuster, 1984.

Brown, Arnold and Weiner, Edith. *Supermanaging: How to Harness Change for Personal and Organizational Success*. New York: McGraw-Hill, 1984.

Buckley, Walter, ed. *Modern Systems Research for the Behavioral Scientist*. Chicago: Aldine, 1968.

Capra, Fritjof. *The Turning Point: Science, Society, and the Rising Culture*. New York: Simon & Schuster, 1982.

Checkland, Peter B. *Systems Thinking, Systems Practice*. New York: Wiley, 1981.

Chubin, Daryl E. et al., eds. *Interdisciplinary Analysis and Research: Theory and Practice of Problem-Focused Research and Development*. Mt. Airy: Lomond, 1986.

Churchman, C. West. *The Design of Inquiring Systems: Basic Concepts of Systems and Organizations*. New York: Basic, 1971.

______. *The Systems Approach*. New York: Dell, 1979.

______. *Thought and Wisdom*. Seaside: Intersystems, 1982.

______, ed. *Systems and Management Annual 1975*. New York: Petrocelli/Charter, 1975.

Clemson, Barry A. *Cybernetics: A New Tool for Management*. Tunbridge Wells: Abacus, 1984.

Cleveland, Harlan. *The Knowledge Executive: Leadership in an Information Society*. New York, Dutton, 1985.

Cook, Norman D. *Stability and Flexibility: An Analysis of Natural Systems*. New York: Pergamon Press, 1980.

Corning, Peter A. *The Synergism Hypothesis: A Theory of Progressive Evolution*. New York: McGraw-Hill, 1983.

Coyle, R. C. *Management Systems Dynamics*. New York: Wiley, 1979.

de Bono, Edward. *Lateral Thinking: Creativity Step by Step*. New York: Harper & Row, 1970.

______. *Six Thinking Hats*. Boston: Little, Brown, 1985.

DeGreene, Kenyon B. *The Adaptive Organization: Anticipation and Management of Crisis*. New York: Wiley, 1982.

de Rosnay, Joel. *The Macroscope: A New World Scientific System*. New York: Harper & Row, 1979.

Doyle, Michael and Straus, David, *How to Make Meetings Work: The New Interaction Method*. New York: Jove, 1976.

Drucker, Peter F. *Managing in Turbulent Times*. New York: Harper & Row, 1980.

Eigen, Manfred and Schuster, Peter. *The Hypercycle: A Principle of Natural Self-Organization*. New York: Springer-Verlag, 1979.

Emery, Fred E. and Trist, Eric L. *Towards a Social Ecology*. New York: Plenum, 1973.

Ferguson, Marilyn. *The Aquarian Conspiracy: Personal and Social Transformation in the 1980s*. Los Angeles: Tarcher, 1980.

Feyerabend, Paul. *Against Method: Outline of an Anarchistic Theory of Knowledge*. London: Verso, 1975.

Fisher, Roger and Ury, William. *Getting to Yes: Negotiating Without Giving In*. New York: Penguin, 1983.

Forrester, Jay W. *Principles of Systems*. Cambridge: Wright-Allen, 1968.

Fuller, Buckminster. *Critical Path*. New York: St. Martin's, 1981.

Gause, Donald C. and Weinberg, Gerald M. *Are Your Lights On? How to Figure Out What the Problem Really Is*. Cambridge: Winthrop, 1982.

Geyer, Felix and van der Zouwen, Johannes, eds. *Sociocybernetic Paradoxes: Observation, Control, and Evolution of Self-Steering Systems*. Beverly Hills, 1986.

Goldsmith, Walter and Clutterbuck, David. *The Winning Streak*. New York: Random House, 1984.

Gollner, Andrew B. *Social Change and Corporate Strategy: The Expanding Role of Public Affairs*. Stamford: Issue Action Publications, 1983.

Goodman, Paul S. et al. *Designing Effective Work Groups*. San Francisco: Jossey-Bass, 1986.

Gray, William and Rizzo, Nicholas, eds. *Unity Through Diversity: A Festschrift for Ludwig von Bertalanffy*. 2 volumes. New York: Gordon & Breach, 1973.

Hall, Arthur D. and Fagen, Richard E. *A Methodology for Systems Engineering*. Princeton: Van Nostrand, 1962.

Hardin, Garrett. *Filters Against Folly: How to Survive Despite Economists, Ecologists, and the Merely Eloquent*. New York: Viking, 1985.

Henderson, Hazel. *Politics of the Solar Age.* New York: Doubleday, 1981.

Hickman, Craig R. and Silva, Michael A. *Creating Excellence: Managing Corporate Culture, Strategy, and Change in the New Age.* New York: New American Library, 1984.

Hoffman, Lynn. *Foundations of Family Therapy: A Conceptual Framework for Systems Change.* New York: Basic, 1981.

Iberall, Arthur S. *Toward A General Science of Viable Systems.* New York: McGraw-Hill, 1972.

Jantsch, Erich. *Design for Evolution: Self-Organization and Planning in the Life of Human Systems.* New York: Braziller, 1975.

______. *The Self-Organizing Universe: Scientific and Human Implications of the Emerging Paradigm of Evolution.* New York: Pergamon, 1980.

______ and Waddington, Conrad H., eds. *Evolution and Consciousness: Human Systems in Transition.* Reading: Addison-Wesley, 1976.

Johnson, David W. and Johnson, Frank P. *Joinging Together: Group Theory and Group Skills.* Englewood Cliffs: Prentice-Hall, 1975.

Jones, J. Christopher. *Design Methods: Seeds of Human Futures.* New York: Wiley, 1980.

Kanter, Rosabeth M. *The Change Masters.* New York: Simon & Schuster, 1983.

Kast, Fremont E. and Rosenzweig, James E. *Organization and Management: A Systems and Contingency Approach.* New York: McGraw-Hill, 1985.

Kauffman, Draper L. *Systems 1.* Minneapolis: Future Systems, 1980.

Kertzner, Harold. *Project Management: A Systems Approach to Planning, Scheduling, and Controlling.* New York: Van Nostrand Reinhold, 1984.

Kets de Vries, Manfred F. R. and Miller, Danny. *The Neurotic Organization.* San Francisco: Jossey-Bass, 1984.

Kilman, Ralph H. *Beyond the Quick Fix.* San Francisco: Jossey-Bass, 1984.

Klir, George J. *Architecture of Systems Problem Solving.* New York: Plenum, 1985.

______, ed. *Applied General Systems Research: Recent Developments and Trends.* New York: Plenum, 1978.

______, ed. *Trends in General Systems Theory.* New York: Wiley, 1972.

Koberg, Dan and Bagnall, Jim. *The Universal Traveler: A Soft-Systems Guide to Creativity, Problem-Solving, and the Process of Reaching Goals.* Los Altos: Kauffman, 1976.

Krippendorf, Klaus, ed. *Communication and Control in Society.* New York: Gordon & Breach, 1979.

Kuhn, Alfred. *The Logic of Social Systems: A Unified, Deductive, System-Based Approach to Social Science.* San Francisco: Jossey-Bass, 1974.

Kuhn, Thomas S. *The Structure of Scientific Revolutions.* Chicago: University of Chicago, 1970.

Land, George T. *Grow or Die: The Unifying Principle of Transformation.* New York: Random House, 1973.

Landa, Lev N. *Instructional Regulation and Control: Cybernetics, Algorithmization, and Heuristics in Education.* Englewood Cliffs: Educational Technology, 1976.

Lasker, George E., ed. *Applied Systems and Cybernetics.* 6 volumes. New York: Pergamon, 1981.

Laszlo, Ervin. *Introduction to Systems Philosophy: Toward a New Paradigm of Contemporary Thought.* New York: Harper & Row, 1972.

______. *Systems Science and World Order: Selected Studies.* New York: Pergamon, 1983.

Lilienfeld, Robert. *The Rise of Systems Theory: An Ideological Analysis.* New York: Wiley, 1978.

Linstone, Harold A. *Multiple Perspectives for Decision-Making: Bridging the Gap Between Analysis and Action.* New York: North Holland, 1984.

Lovelock, James E. *Gaia: A New Look at Life on Earth.* New York: Oxford University, 1979.

Machlup, Fritz and Mansfield, Una, eds. *The Study of Information: Interdisciplinary Messages.* New York: Wiley, 1983.

Maier, Norman R. H. et al. *Team Effectiveness and Meeting Strategy.* Plymouth: Human Synergistics, 1977.

Majone, Giandomenico and Quade, Edward S., eds. *Pitfalls of Analysis.* New York: Wiley, 1980.

Mason, Richard O. and Mitroff, Ian I. *Challenging Strategic Planning Assumptions.* New York: Wiley, 1981.

Maturana, Humberto R. and Varela, Francisco J. *Autopoiesis and Cognition: The Realization of the Living.* Boston: Reidel, 1980.

Meadows, Donella H. et al., eds. *Groping in the Dark: The First Decade of Global Modelling.* New York: Wiley, 1982.

Miles, Ralph F. *Systems Concepts.* New York: Wiley, 1973.

Miller, James G. *Living Systems.* New York: McGraw-Hill, 1978.

Miser, Hugh J. and Quade, Edward S., eds. *Handbook of Systems Analysis.* New York: North Holland, 1985.

Morgan, Gareth. *Images of Organization.* Beverly Hills: Sage, 1986.

______, ed. *Beyond Method: Strategies for Social Research.* Beverly Hills: 1983.

Mouton, Jane S. and Blake, Robert R. *Synergogy.* San Francisco: Jossey-Bass, 1984.

Myers, Norman, ed. *Gaia: An Atlas of Planet Management.* New York: Doubleday, 1984.

Odum, Howard T. *Systems Ecology: An Introduction.* New York: Wiley, 1983.

Olsen, Shirley A., ed. *Group Planning and Problem-Solving Methods in Engineering Management.* New York: Wiley, 1982.

Pask, Gordon. *Conversation, Cognition, and Learning.* New York: Elsevier, 1976.

Pastin, Mark. *The Hard Problems of Management: Gaining the Ethics Edge.* San Francisco: Jossey-Bass, 1986.

Pedler, Mike, ed. *Action Learning in Practice.* Aldershot: Gower, 1983.

Peters, Thomas J. and Waterman, Robert H. *In Search of Excellence: Lessons From America's Best-Run Companies.* New York: Harper & Row, 1982.

Piaget, Jean. *The Construction of Reality in the Child.* New York: Basic, 1959.

Pinchot, Gifford. *Intrapreneuring: Why You Don't Have to Leave the Corporation to Become an Entrepreneur.* New York: Harper & Row, 1985.

Pirsig, Robert M. *Zen and the Art of Motorcycle Maintenance: An Inquiry Into Values.* New York: Bantam, 1975.

Powers, William T. *Behavior: The Control of Perception.* Chicago: Aldine, 1973.

Prigogine, Ilya and Stengers, Isabelle. *Order Out of Chaos: Man's New Dialogue With Nature.* New York: Bantam, 1984.

Quade, Edward S. *Analysis for Public Decisions.* New York: North Holland, 1982.

Rapoport, Anatol. *General System Theory: Essential Concepts and Applications.* Tunbridge Wells: Abacus, 1986.

Reckmeyer, William J. *The Emerging Systems Paradigm: An Historical Perspective.* Ann Arbor: University Microfilms, 1982.

______. Invited address on "Management Cybernetics: Theory and Praxis in the Systems Age" at the University of Stockholm and the Royal Institute of Technology. Stockholm: 1984.

Richardson, Jacques, ed. *Models of Reality: Shaping Thought and Action.* Mt. Airy: Lomond, 1984.

Robinson, Mike. *Groups.* New York: Wiley, 1984.

Sage, Andrew P. *Methodology for Large-Scale Systems.* New York: McGraw-Hill, 1977.

Schein, Edgar H. *Organizational Culture and Leadership.* San Francisco: Jossey-Bass, 1985.

Schieve, William S. and Allen, Peter M., eds. *Self-Organization in Dissipative Structures.* Austin: University of Texas, 1981.

Schon, Donald A. *The Reflective Practitioner: How Professionals Think in Action.* New York: Basic, 1983.

Segal, Lynn. *The Dream of Reality: Heinz von Foerster's Constructivism.* New York: Norton, 1986.

Shanin, Teodor, ed. *Rules of the Game.* London: Tavistock, 1972.

Sheldrake, Rupert. *A New Science of Life: The Hypothesis of Formative Causation.* London: Blond & Briggs, 1981.

Swap, Walter C. et al, eds. *Group Decision Making.* Beverly Hills: Sage, 1984.

Tichy, Noel M. *Managing Strategic Change: Technical, Political, and Cultural Dynamics.* New York: Wiley, 1983.

Toffler, Alvin. *The Adaptive Organization.* New York: McGraw-Hill, 1985.

_____. *The Third Wave*. New York: Morrow, 1980.

Trappl, Robert ed. *Cybernetics: Theory and Applications*. New York: Hemisphere, 1983.

______, ed. *Power, Autonomy, Utopia: New Approaches to Complex Systems*. New York: Plenum, 1986.

van Gigch, John P. *Applied General Systems Theory*. New York: Harper & Row, 1978.

______, ed. *Decision Making About Decision Making: Metamodels and Metasystems*. Tunbridge Wells: Abacus, 1987.

Varela, Francisco J. *Principles of Biological Autonomy*. New York: North Holland, 1979.

Vickers, Geoffrey. *The Art of Judgement: A Study of Policy-Making*. London: Chapman & Hall, 1965.

______. *Freedom in a Rocking Boat: Changing Values in an Unstable Society*. Baltimore: Penguin, 1972.

von Bertalannfy, Ludwig. *General System Theory*. New York: Braziller, 1968.

von Foerster, Heinz. *Observing Systems*. Seaside: Intersystems, 1982.

______ and Zopf, George P., eds. *Principles of Self-Organization*. New York: 1962.

______ et al., eds. *Cybernetics of Cybernetics: Or the Control of Control and the Communication of Communication*. Urbana: Biological Computer Laboratory—University of Illinois, 1974.

von Oech, Roger. *A Kick in the Seat of the Pants: Using Your Explorer, Artist, Judge, and Warrior to be More Creative*. New York: Harper & Row, 1986.

______. *A Whack on the Side of the Head: How to Unlock Your Mind for Innovation*. Menlo Park: Creative Think, 1982.

Waddington, Conrad H. *Tools for Thought: How to Understand and Apply the Latest Scientific Techniques of Problem-Solving*. New York: Basic, 1977.

Warfield, John N. "Organizations and Systems Learning." *General Systems* 27 (1982): 5-74.

______ et al. *Generic Design*. Fairfax: Center for Interactive Management—George Mason University, 1986.

Watzlawick, Paul, ed. *The Invented Reality: How Do We Know What We Believe We Know? Contributions to Constructivism*. New York: Norton, 1984.

Watzlawick, Paul et al. *Change: Principles of Problem Formulation and Problem Resolution*. New York: Norton, 1974.

Weinberg, Gerald M. *An Introduction to General Systems Thinking*. New York: Wiley, 1975.

______. *The Secrets of Consulting*. New York: Dorset, 1985.

Weinberg, Gerald M. and Weinberg, Daniela. *On the Design of Stable Systems*. New York: Wiley, 1979.

Wiener, Norbert. *Cybernetics: Or Control and Communication in the Animal and the Machine*. Cambridge: M.I.T., 1949.

Wilden, Anthony. *System and Structure: Essays in Communication and Exchange*. London: Tavistock, 1980.

Yovits, Marshall C. and Cameron, Scott, eds. *Self-Organizing Systems*. New York: Pergamon, 1960.

Yovits, Marshall C. et al., eds. *Self-Organizing Systems 1962*. New York: Pergamon, 1962.

Zeleny, Milan. *Multiple Criteria Decision Making*. New York: McGraw-Hill, 1982.

______, ed. *Autopoiesis: A Theory of Living Organization*. New York: North Holland, 1981.

DISCUSSION OF RECKMEYER'S PRESENTATION

Sadovsky

1) A traditional question: You have used the term "metaparadigm." Explain to us, please, the meaning of this term? Ordinarily we use the term "meta..." in two distinct senses: (a) as a designation of a theory in the framework of which the corresponding theories are described; (b) as a designation of a more general rule, law, theory, etc. What is your meaning?

2) A delicate question: In the very interesting bibliography which you kindly present to all of us, I do not find any items of relevant Russian literature (of course, in English translation). Are there some reasons for this?

Reckmeyer

1) I think both apply—the systems approach is a metaparadigm that consists of very general insights and tools (your second interpretation). It is a framework including more specific insights and tools—systemic and disciplinary—whatever their nature (your first interpretation), including fundamental epistemological, ontological, and methodological features.

2) Your "delicate question" is fairly easily answered. Except for one or two Russian texts in English translation that I know of (none of which I think really address management concerns), I do not know the Russian literature very well. It would help us here in the United States if you and your colleagues would keep us informed about what is available in English translation—we would really like to incorporate them into our work on a more regular basis.

Umpleby

Bill, why not simply acknowledge that the USSR is a more open society? (smiles and chuckles)

Von Glasersfeld

We need to organize translations of key material.

Zeleny

Consider the case of a dictator who "successfully" managed a country in that he stayed in power. Did he have requisite variety? Dampening variety is a management approach. But there are also variety expanding possibilities in self-organizing systems.

The law of requisite variety appears *not* to be applicable to social systems but only to machines and mechanistic systems. The one-to-one "matching" of variety does not take into account any self-organizing, spontaneous, creative, innovative, autonomous, problem-solving and other aspects of *human* initiative. None of this appears in Ashby's law.

One can manage most effectively not by matching human systems variety but by "bowing" to that variety and by allowing it to manifest itself in full freedom and respect. This approach does not require requisite variety on the part of the "regulator." Do you object?

Reckmeyer

I think that Ashby's notion—as broadly interpreted—is relevant to *any* kind of system. There are numerous ways in which one could address your questions, all too extensive to elaborate here. The essentials, it seems to me, include the following points.

First, Ashby talks primarily in terms of how to regulate—that is, any regulator must be at least as complex as the system to be regulated; my view extends Ashby's law to systems in general and doesn't presuppose the case of regulation.

Second, environments are specific to systems, not a general context that is identical for all systems; as such, any living system that lasts for a significant period of time clearly has requisite variety in order to do so.

Third, the key is how to expand one's environment by amplifying one's own variety rather than dampening it. We all exist by having to dampen variety to some extent; the problem is the degree, my view being that people generally dampen external variety because it is much easier to do so.

Fourth, the most effective and most ethical way to amplify variety in the human domain is to move away from the nervous system-oriented view of regulation to a more ecologically-oriented one that emphasizes the multi-dimensional, self-organizing, autonomous nature of human activity systems—individuals, families, organizations, communities, nation-states, etc. In so doing, it seems to be imperative to recognize that regulation is a *relational process*, not a state of affairs or structured part in a system. Regulation involves at least two (or more) parties in *mutual* interaction; as such, the central question is: which regulates which and to what extent. The conventional view is that one regulates another; my view is that both participate in the regulatory process and that we need to move away from the conventional hierarchical view of linear control to this more useful view of heterarchical mutual control—all of which is explicitly derived from and consistent with the points you raise.

With my family I amplify their autonomy to the limits of my patience.

Rapoport

When you use these methods, do you have the feeling that responsibility is lost? I must be responsible for what I propose or I shall be considered a poor adviser.

Reckmeyer

I have found that responsibility in all participants goes up. I am concerned with trying to tap the wisdom and experience of the group.

Umpleby

Bill's comment that he tries to amplify autonomy in his family to the limits of his patience is in contrast to child rearing practices in the Soviet Union. Soviet children are smothered with love and attention and learn to feel nurtured by a complex web of rules, cautions, and restraints, according to New York Times correspondent David Shipler.

CHAPTER

13

THE ACTIVITY-SYSTEMS APPROACH TO DEVELOPMENT OF THE HUMAN FACTOR IN INNOVATION

Nickolai I. Lapin and Boris W. Sazonov

ABSTRACT

This paper deals with the active nature and inner contradictions of the human factor in innovation. New trends and methods for increasing the rate of innovation are described.

Introducing radical innovations, for example, the mass production of "slit light guides," showed that the first structures of innovative activity are half-formed and even non-formal in their nature. They are based on the enthusiasm, creativity and other interests of the participants. It is asserted that the forecast of success and control of innovations demands discovering stable organizational and "situational" structures which come into being in the course of an innovation process within the framework of these participants' "positions" (such as initiative, co-operation, passivity, opposition). These "positions" are defined by two factors: the system of indicators accepted in the organization introducing an innovation and the psychological attitudes of individuals and groups to the innovation.

Social-organizational innovation is examined as a combination of new methods for development of the human factor in innovation. The methodology of "innovation games" is proposed which enables simulation of real innovation processes while "reducing" them in space and time, in order to discover sooner the disadvantages of an innovation project, the potential conflicts caused among groups of people by a project, and ways of resolving the conflicts.

Innovation is a complete dynamic system including many heterogeneous factors: technology, organizational structure, and human interests. These factors were described earlier [1].

In the present report, we shall focus on the human factor: its nature as an activity and inherent contradictions, also new trends and methods of its use and development.

1 THE HUMAN FACTOR IN INNOVATION: ACTIVE AND CONTRADICTORY NATURE

Viewed from the point of view of general theory, the core of the innovation process consists of innovation activities. In other words, innovation is a method used to arrange innovative activities [2]. This method or organizational form is systematic in its nature. It is an inherently integral entity which is dynamic and interacts with its environment (social and material). However, a scientific investigation of an innovation is also a kind of innovative activity which, on the one hand, reflects and analytically reproduces its object. On the other hand, it develops a set of tools which provide for the necessary progress in innovation. These are the original concepts of the "activity system" approach to innovation developed by the authors [3,4].

Within the framework of this concept, the "human factor" in innovation is nothing but innovative activities performed by individuals and social groups acting as immediate subjects of innovation processes (other than the activities carried out by institutionalized subjects, that is, organizations). Innovative activity as such is a process in which people produce new products and methods for their manufacture. It is directed toward changing the components of routine, productive activities. It follows that human innovation activities contradict habitual productive activities, and hence negate them. This negation occurs in two forms: radical and modifying, or improving forms.

These forms represent two basically different types of innovation activities and are characterized by significantly dissimilar mechanisms. While their interrelation is dialectical in nature, following introduction of a radical innovation, improving activities promote its further progress and prolong its effects. At the earlier stages, a radical innovation encounters resistance from the mechanisms and structures involved, not only in routine but also in improving activities. Therefore, at the initial stages of a radical innovation, the role of the human factor is especially important.

This statement can be illustrated by an example of the manufacture of lighting equipment for domestic and industrial purposes. Research testified to the fact that the organizational structures supporting the manufacture of lighting equipment primarily correspond to the mechanisms of productive and improving activities. Radical changes or innovative activities stand aside as if they were something unique needing specific organizational support. Consequently, they have been set up anew for each particular case.

The mechanisms of improving activities require that each product in need of improvement is treated as a set of components, each of which is modified relatively separately in accordance with specific logic and criteria. Thus, sources of light are improved according to criteria of efficiency, longevity, cost reduction, etc. Lamp fittings and frames are improved in shapes, manufacturing technologies, saving of scarce materials, etc. Each component is improved and provided with individual scientific/technical, organizational and production support. As a result, improving activities tend to be a routine process. Here the human factor plays a role similar to the one in regular routine production practices.

The mechanism of drastic innovative activities is quite different. It aims at producing a qualitatively new product rather than improving the old one part by part. An innovation in lamp making can be illustrated by the development of launching into mass production of slit light guides. A slit light guide is a relatively thin but sufficiently long (up to several dozen meters) hollow tube in which the inner surface is coated with light-reflecting film. A source of light placed at one end of the tube provides light which is reflected again and again and finally emerges through a slit of certain dimensions [5]. Slit light guides are particularly efficient in explosive environments since the source of light can be placed outside the lighted space. In large industrial facilities slit light guides contribute to savings in the maintenance of lighting equipment. Designers are offered new opportunities for improved design of interior and exterior architecture and the arrangement of space.

The idea of a slit light guide can be traced back to 1874-1878 when V. Chikolev, a Russian engineer, tried to use it in the lighting system of the Petersburg powder plant. However, the first model proved inadequate, and the idea was forgotten for quite a long time. Ninety years later, in 1963, the idea of slit light guides was revitalized by G. B. Bukhman, a specialist employed at a construction engineering design office having little to do with lighting systems. Having initiated a development program, G. B. Bukhman was supported by the staff of the All-Union Lighting Engineering Research Institute, primarily by Y. V. Isenberg and V. M. Pyatigorsky. In 1972 development of slit light guides was included in plans of the State Committee for Science and Technology, the above-mentioned Institute, and an enterprise which was to manufacture them (Vatra Combine). In 1975 mass production began.

Development of slit light guides proved to be a complex process involving several industries. It required research, technical invention, and drastic transformation of the work patterns at some enterprises of electrical engineering and in some chemical industries. Major difficulties were encountered in producing a high-quality light-reflecting film and in developing a technology for its adequate adhesion to the inner surface of the tube.

The success of this cardinal innovation was greatly facilitated by the outstanding personality of the initiator and leading designers. These specialists employed in different industries established an informal working relationship, built on a common desire to bring their work to a successful end. Initially, the design of the slit light guide was made within the framework of a semi-formal structure set up for this purpose in a design office specializing in other than light engineering problems. In the beginning and later stages of the process, informal contracts characterized the development and manufacture of many lamp components and servicing systems. There were always enthusiastic people ready to face unusual tasks. It was later that their results were consolidated by plans or agreements on creative cooperation. The participants in this innovation process were motivated by the continuation of their own creative pursuits with new materials, an opportunity to prepare a dissertation thesis, or some "pay" for their participation in the work of a creative team.

Thus at the initial stages of a radical innovation process, the role played by the human factor proves decisive. As interaction among the participants in the process becomes more and more formalized, new and more powerful organizational and economic stimuli step in to back up the human factor, which will never fade away altogether. Moreover, at some stages of process formalization it may again grow in importance. For instance, if formalization proceeds from contradictory information, the human factor becomes a hindrance to the innovation process.

Research carried out by our colleague, A. I. Prigozhin [6] shows that groups participating in an innovation process can assume different, often contradicting approaches to their roles in this process. They can be full of initiative or cooperation, or else be idle or even stand in opposition. The choice of a position in their attitudes is determined by two interrelated factors: the system for evaluating the work of organizations participating in the innovative process and the psychological attitude of individuals and groups to novelties. Investigation into a radical novelty showed that an initiative stance was taken primarily by the development staff. Cooperation was characteristic of managers, designers and users, however, the latter were often idle. Manufacturers frequently stood in opposition to changes in production processes when the innovation lacked adequate economic incentives.

The research results obtained by our colleagues V. S. Tolstoy [7] and V. S. Dudchenko [8] also showed that the multiplicity of roles and organizations associated with an innovation process demands that specific structures of the participants' relationships be taken into account. Two types of structures—predictive and situational ones—have been identified and analyzed. Being drastically different in content, they perform co-relative functions in the innovation process. These are partly incompatible and partly complementing.

Predictive relationships are directive. They are obligatory for those concerned. They include not only formal relationships but also those spontaneous, informal relationships which motivate persons to stick to respective programs of action. To bring about an innovation, it is necessary to introduce some transformation in predictive relationships, both in volume and in structure. However, it does not always work, and then the innovation process slows down and some other hindrances emerge. Invariability is primarily characteristic of relationships between organizations. Within organizations it is typical of relationships between the administration and employees as individuals.

Meanwhile, it has been found that due to their inherent logic and creative potentialities, innovation processes encourage setting up specific situational structures in relationships among the participants in innovations, both within and between organizations. A situational structure is an integral set of individuals' attitudes, expectations, and actions directly caused by an innovation situation, which finally represents the operation of basic underlying factors. These factors are the nature of the balance in the organization among specific latent structures and mechanisms. In some cases, situational structures promote innovations by compensating for unavailable predictive structures, or by providing for further development of innovation processes bypassing obsolete structures of this type. In other cases, they hinder innovations by supporting non-innovative stationary structures or by conflicting innovation structures.

Analyses in the active nature and contradictions of each human factor in innovation testify to the urgency of the search for new directions and methods of its utilization and development.

2 THE HUMAN FACTOR OF INNOVATION: ELABORATION OF NEW LINES AND METHODS OF DEVELOPMENT

We shall deal with two of a broad variety of potentially possible lines and methods of promoting the human factor in innovation: socio-organizational design of innovation and innovation games.

Research results show that it is not enough to design only technical and economic parameters for the production of novelties, the more so when confined to a single organization. It is necessary to perform a truly comprehensive design of innovation processes, including their socio-organizational parameters, both for internal management and inter-organizational relations.

A socio-organizational innovation project is in fact a system of social and organizational variables which describe the structure of the novelty of its dynamics as a complex process in which various groups of people participate and cooperate with each other at its different stages. Design should involve such groups of parameters as: organizational-structural ones (organizations participating in the innovation, their subdivisions, predictive relations, etc.), socio-structural ones (professional and skills characteristics, socio-demographic groups, etc.), functional-process factors (aims, stimuli, etc.) and others.

The designed parameter values are to proceed from some standards and analyses of concrete organizations participating in innovation. The purposes of the system of social and organizational parameters of designed innovation are to incorporate the main objective of the innovation and subordinate tasks, be capable of ascertaining the major socio-organizational problems which might hinder realization of the innovation, and formulate the basic tasks to be handled in solving the problems and the lines along which these problems are to be handled [9].

At the conceptual design stage, the main social aim of the innovation is formulated and the organizational and socio-professional structure of partner groups is stated and analyzed. At the contractor design stage, detailed analyses are performed of the subordinate tasks, including all parameters, problems, and means to be used in handling them at all major stages of the innovation process. Meanwhile, alongside the parameters of internal organization, special emphasis should be placed on those of relationships between organizations engaged in innovation, both structural and dynamic, so that maximum cooperation can be secured for the aims and interests and attitudes of innovation partners employed in different organizations. In other words, the aim is to secure the highest possible level of integration in the innovation process.

Comprehensive inter-organizational innovations also need task-oriented programming or cut-across planning as a complex process (from development to implementation). It must be clear who participates in their support and what particular measures will be taken to promote and control them to achieve maximum perfection and efficiency.

Specialized sociological services employing a broad range of methods and procedures are highly useful in developing socio-organizational projects and innovation programs, to say nothing about their implementation. Among these methods we distinguish innovation games, a new type of business game developed with the active participation of the authors and their colleagues. They are advantageous in that they make possible simultaneous handling of three interrelated tasks: shaping new methods of solving applied problems, scientific investigation of the characteristic features of the human factor, and practical promotion of all the characteristic features.

The methodological core of an innovation game is the activity approach which has been developed during the last decade on the basis of materialistic dialectics. Due to this approach an innovation game can be regarded as an analogue of real innovation processes described above in their active and contradictory nature and as a means of affecting them in a practical way.

The techniques and procedures of an innovation game make use of some elements of Lewin-Skein's model "unfreezing—change (action)—freezing" [10], and of Tavistok's "living methodology" aimed at "revealing role boundaries," and releasing human abilities for innovation [11].

An innovation game has a rather complicated structure. Thus, the program of one of its versions includes eight interrelated and, to a considerable extent, concurrently exercised subprograms: decision making, design, programming, education, training, recreation, research, organization [12].

An innovation game makes it possible to reproduce real innovative processes "compressing" them in time and space while ascertaining the deficiencies of an innovation project and the probable conflicts between the parties concerning the project and its implementation. This is done by facilitating the improvement of the project and the program for an innovation activity, by revealing diverse consequences of novelty and by finding ways of overcoming conflicts. As participants take part in the game, they prepare draft instructions for themselves. Therefore the process is self-organizing.

These and other functions of the innovation game are performed because, as it proceeds, real cooperation and conflict occur among the potential participants. The directors purposefully assist the participants in "disclosing" many facets of their social being. In these various situations, particularly in the course of conflicts and attempts to overcome them, people expose what was concealed from both participants and researchers, and it becomes evident that a phenomenon which seemed "monoatomic" is, in fact, complex.

Innovation games enable one to distinguish the lines of behavior of the participants, which may be determined by significantly dissimilar factors. First, there are relatively stable factors which are independent of the game proper. These are environmentally shaped. They are usually not realized socio-cultural objectives, habitual attitudes, occupations, or assumed roles. Second, there emerge situational factors as the game goes on including attitudes towards the game as a whole and its elements and towards the interactions in the game.

These facets are ascertained as a result of both the spontaneous activities of the participants in the game and the purposeful work of its organizers. The organizers arrange the course of the game and its structure in order to achieve pre-set goals and to eliminate deadlock situations which might emerge during the game. Thus an innovation game can be treated as a form of social experimentation. A few mutually complementary techniques of such experimentation have been developed. Each allows us to ascertain a unique set of features characteristic of the human innovation factor (making them an object of control at the same time).

1. The game participants are classified by their social features. Among such features the most evident and relevant are formal roles (occupational positions) played by the participants prior to their incorporation into the innovation process (or the game which simulates it) and professions (trades). The manager of a game may set up several monorole (monoprofessional) groups to prepare a single innovation project. In this case the inter-role (monoprofessional) contradictions emerging in the course of the innovation process prove to be the major motive mechanism. Conflicts emerging within such groups are primarily caused by different attitudes. They are analyzed to account for both relatively stable life stands and situational attitudes arising in the course of the game, which can also be intricate and dissimilar. Such attitudes are those towards managers and the participants in the game, the game as such, the objectives of the game and of the innovation, etc.

At the level of inter-group interaction, it becomes clear that while simulating an innovation process specific roles are shaped: those of an innovator, organizer, designer, programmer of the innovation process, critic, etc. It often happens that a purely formal role assumed by a participant prior to the game (innovation process) becomes his role in this process (autonomic—incorporations of the role). Such roles include those which perform a service function relative to the innovation: budgeting, marketing, manufacturing, etc. Meanwhile, the personal attitude towards the innovation adopted by a person who plays such a role will often be determined by the organizational structure in which this role exists rather than by the life stand of this person.[1]

---

[1]Investigation into actual innovations shows that even if one's life stand comes to the forefront, a probable contradiction between the role and the stand is eliminated in creative management which may be formal in an attempt to reconstruct an existing organizational pattern, or informal in building up an informal structure alongside a formal one. This phenomenon is hardly discernible in a game. The above example illustrates one of the many restrictions imposed by research carried on in the form of a game.

Different results occur in games in which the participants are divided into groups, each including a complete set of innovation roles (professions and trades involved in the innovation). In these games each group prepares its own self-sufficient project in order that the best one can be selected as a result of their competition. In such cases, the best mechanism for the solution of innovation problems is found in inter-group competition. Processes of intra and inter-group growth, which are the subject of socio-psychological research, come to the fore. Even though this organizational form is sufficiently effective for handling innovation tasks, it is hardly fruitful as a method for investigating the human factor in innovation since many inter-role and inter-professional conflicts significant in innovation tend to be secondary compared to the situational game and life stands taken by the participants.

2. Reflection on the role and stands of the game participants is arranged. Reflection on all situations formed during the game is inevitable in any innovation game, particularly in cases when a deadlock arises or it becomes necessary to investigate the nature of conflicts. Reflection dismantles obstacles to further progress. However, those who arrange the game may take the next step by transforming the results of reflective activities into a specific semiotic operative system which makes it possible to arrange joint thinking of the participants in the game and to arrive at positive solutions concerning the problems emerging at subsequent stages of innovation. A similar technique was proposed in the so-called "organization activity" game [13]. It basically consists of the following procedure.

Conflicts arising among the participants in a game are appraised in view of their socio-cultural significance. If they are significant, that is, if a conflict observed in the game involves more than a personal disagreement, then the managers of the game treat the simulated situation as an example to be used in searching for ways of settling the conflict. This search is performed by constructing new activity structures that eliminate the conflict rather than sacrifice the interests of one of the conflicting parties in favor of the other.

This technique not only contributes to the solution of practical problems hindering a novelty but also proves to be useful in cognizing (self-knowledge) and developing (self-evolution) numerous aspects—primarily in the socio-cultural sphere—of the human factor in innovation.

3. Separate games on issues which arise in the course of the preceding games are sometimes arranged. An innovation game aims at meeting a practical goal by simulating the whole or part of a particular innovation process. However, performance of some component of an innovation activity is frequently hindered, not only because the innovation is complicated or the environment rejects it, but also because there are no standards for the activity, or the current standards are inadequate for the new case. Thus, games give rise to questions about how the innovation problem is to be formulated, what the innovation project and program should be like, how communication should be arranged in the course of an innovation game, what is a socio-cultural conflict, and with what means it can be settled, etc.

An important feature of games is that these and other questions can be answered using the means offered by the game. For instance, standards for an innovation game can be not only "imposed" by a theoretician of design activities but also formulated in the course of a game. To do so, a goal is set (in our case, to prepare an innovation project) for which there are presently no socially acceptable solutions. Participants try to create a solution by reflecting on the process of problem solving and by translating the methods found by them into standards for respective (in our case, design) activities. This method relies on the belief that joint effort, supported by corresponding methodological services (which arrange reflection, etc.), is capable of creative results. Like activity standards which are not invariable forever but steadily develop with respective activities, these methodologically-oriented games are repeated time and again, which is one of the factors governing progress on such standards.

To sum up, one can assume that the above techniques used in arranging innovation games act simultaneously as a method handling specific practical tasks, for doing research and for developing the human component of innovation processes. In order to deal with this triple task, it is necessary to have well-trained managers and advisors capable of distinguishing these three tasks and offering ready solutions or developing new ones jointly with the participants in the game.

REFERENCES

1. Lapin, N. I. Problems of Systematic Research on Innovations. In: *Management of Scientific and Technological Progress: Successes and Some Perspectives.* Moscow, s. 28, MCSTI, 1982. (In Russian).
2. Lapin, N. I. Intensification of Innovation Processes: a Strategic Goal of Innovation Theory and Practice. In: *Innovation Processes.* Proceedings of a Seminar. Moscow, Central Systems Research Institute, 1982. (In Russian).
3. Lapin, N. I., Prigozhin, A., Szonov, B., Tolstoy, V. Innovations in Organizations. In: *The Structure of an Innovation Process.* Proceedings of a Conference, Moscow, Central Systems Research Institute, 1981. (In Russian).
4. Sazonov, B. Activity Approach to Innovation. In: *The Social Factors of Innovation in Managerial Systems.* Proceedings of a Conference, Moscow, Central Systems Research Institute, 1980. (In Russian).
5. Knorring, G. Lighting Plant. Subsection 2.11 "Slit Light Guides as Lamps", Moscow, Energoizdat, 1981. (In Russian).
6. Prigozhin, A. Inter-Organizational Role Positions in an Innovation Process. In: *The Structure of an Innovation Process.* Proceedings of a Conference, Moscow, Central Systems Research Institute, 1981. (In Russian).
7. Tolstoy, V. Acceleration of an Innovation Process by Designing Predictive Relations. In: *Innovation Processes.* Proceedings of a Seminar, Moscow, Central Systems Research Institute, 1982. (In Russian).
8. Dudchenko, V. Dynamics of Situational Structures in an Innovation Process. In: *Innovation Processes.* Proceedings of a Seminar, Moscow, Central Systems Research Institute, 1982. (In Russian).
9. Lapin, N. Urgent Problems of Improving Social Design Under Present Conditions. In: Toschenko, J., Aitov, N., Lapin, N. *Social Design*, Chapter Six, Moscow, Mysl Publishers, 1982. (In Russian).
10. Lewin, K. Group Decisions and Social Change. In: *Readings in Social Psychology*, Maccoby, Newcomb and Hartly (eds.), Holt, Rinehart and Winston, N.Y., 1947.
11. Lawrence, W. Gordon (ed.) *Exploring Individual and Organizational Boundaries: A Tavistok Open Systems Approach*, Chichester - N.T., 1979.

12. Dudchenko, V. et al. A Program for an Innovation Game, Jaroslavl, Institute for Refresher Courses for Specialists in Petrochemistry, 1982. (In Russian).

13. Shchedrovitsky, C. Automation in Design and the Goals of Promoting Design Activities. In Collected Papers, *Development and Introduction of Automated Systems in Design Work: Theory and Methodology*, Moscow, Stroiizdat, 1975. (In Russian).

DISCUSSION OF LAPIN'S PRESENTATION

Umpleby

Where are these methods being used?

Lapin

In advanced business schools. One institution adopted the methods. A commission came to investigate. Initially they were negative, because one condition of the method is—no drinking. One week later twelve members of the commission voted for the methods, one person was opposed, and one abstained. All managers who go through that training program will use this method—the Academy of the National Economy, an advanced business school for managers including ministers.

Sometimes one has a successful outcome, with audience support, but the teachers object.

Lavrov

What Lapin says is correct. A short time ago I had to prepare a report for our government on the work of the advanced training institutes in our country. It is interesting now to note that in our report we (i.e., a group from the Academy of the National Economy, the Ministry of Higher Education and other organizations) urged a considerable expansion of the practice of "business games".

Lapin

These methods are the result of ten years work at the Institute for Systems Studies. We have been issuing reports in Russian.

There is a problem of intensifying and diagnosing. We know Stanley Wenk at the National Academy of Management, Thomas Moss and his report to Congress, other Western specialists. I also gave a report to the International Institute for Applied Systems Analysis in Vienna, Austria, and to businessmen from the U.S. We have connections to the Tavistock Institute through Professor Gordon Lawrence.

Reckmeyer

1) Are the "innovation issues" used in your work formulated by the facilitators or do they come from the real-life experiences of the game's participants?

2) You mentioned your major concern right now is "how to transfer these results to other groups"—does this mean a difficulty in transferring the insights and skills themselves (a pedagogical concern) or a difficulty in enabling clients to make this "innovation approach" operational in their own work after the game is over—or both?

Lapin

1) The task of the organizers of the game was not to suggest decisions but to create the conditions for creative decision making by all participants, to stimulate their activity.

2) There are pedagogic and bureaucratic obstacles to the transfer of these methods and others as well. But we also have had some successful experiences. For instance, the method of innovation games will take as its basic task that of heightening managerial skills (a la business school). This method was approved by the Academy of the National Economy where the higher-level managers are trained. But this is just the beginning—even though it gives cause for optimism.

Mitroff

Do you use a dialectical method—two or more different groups operating with different definitions of constraints?

Rapoport

It is impossible to form equal teams (from the point of view of their creative potential and competence) for dialectical competition on decisions. We use a method where a two-level structure is formed: the basic level draws from participants working on the general task and the higher-level draws the most competent participants with the highest responsibility for the concrete tasks.

The higher level team considers all alternatives raised by the basic team and reconstructs them from the point of view of the concrete task using more confidential criteria which may exist for a firm's activity in the particular situation.

This method has some dialectical essence too. What do you think of it?

Mitroff

The dialectrical method is a good way to surface assumptions. Methods often stress participation, but the boss still must decide. We have been able to form teams that are approximately equal.

Lapin

In each game we have between thirty and forty participants. We need three groups. More than four groups is too many. People choose their groups and can transfer to another group.

At vespers participants can criticize the organizers. These discussions often become heated.

The games are intended to: 1) solve practical problems, 2) involve people in implementation, and 3) generate new material for scientific investigation. We analyze the events that take place as an object of study.

Mitroff

Group composition is important. By creating a group in a particular way one can predict what the group will produce.

Zeleny

The atmosphere created in your game (freedom, voluntarism, criticism) is different and incompatible with the actual conditions in companies and plants. Is the purpose to show people their potential and then return them into conditions where these new potentials cannot be realized? Or, is the purpose to change the creativity-suppressing conditions which prevail in actual firms?

Lapin

This is an accurate and difficult question. There does indeed arise a contradiction between creative decisions arising as a result of the game and the limited opportunities in actual organizations. This contradiction must be taken into consideration in the draft decision which is elaborated in the course of the game, i.e., it must reveal ways of overcoming the contradictions noted. It is also important not to lose time and to use the solidarity and determination of participants in the game to achieve changes, i.e., to create the conditions for implementing the draft solution to the problem. It is also important to repeat the game with the same participants (of course, on a different theme) about 6 months later so that the group can continue to exist, and then make innovation games a regular element in the normal development of the given organization.

Mitroff

If people are allowed to play the game but not to implement the methods in their firms, they will only participate once.

Zeleny

If the existing organizational and bureaucratic limits cannot be changed, isn't it more productive to focus the game on human functioning in real environments rather than creating "ideal" conditions which could be more harmful to the personalities of the participants?

Lapin

Of course the game should lead not to some "ideal" change but to one which is realizable. One of the major tasks of the game is to explore a plan for implementing such changes. Without such a "metaplan" or plan for organizational change, the game cannot be considered complete.

CHAPTER

14

THE SYSTEMS APPROACH TO INTERNATIONAL RELATIONS: A COMPARISON OF U.S. AND SOVIET MODELS

Bruce J. Allyn

In the Soviet Union beginning in the 1960s, systems analysis became a subject of tremendous official enthusiasm and support. Beyond its numerous military applications, systems theory and analysis was applied in Soviet society under the party slogan of "scientific management of society" (*nauchnoe upravlenie obshchestvom*)[1] with special stress on areas such as the development of automated management systems.[2] With time the systems "approach" began to be applied to international relations research. The 1970s saw the establishment of the Center for Systems Analysis of International Relations at the Moscow State Institute of International Relations (MGIMO) under the Foreign Ministry. Soviet researchers also began to model international trade and politics at the All-Union Institute for System Research (VNIISI), which is jointly subordinate to the State Committee on Science and Technology and the Soviet Academy of Sciences.

1 SOVIET SYSTEMS RESEARCH OF INTERNATIONAL RELATIONS

In general, Soviet writings treat systems analysis and the systems approach as a particular method, which is not in competition with Marxism-Leninism as a theoretical framework for understanding society or international relations. Soviet writing defines historical materialism as a philosophical science of the general laws of social development; particular aspects of social life are the objects of research by various scientific methods, system analysis being one of them.[3] As a 1978 Soviet publication reads:

> It may be said that systems analysis stands in the same relation to materialist dialectics as a particular theory to general theory. The methodological function of systems analysis is usually confined to problems of separate processes and practical tasks. The dialectical method, on the other hand, uncovers universal laws of the process of development. This difference between the systems approach and dialectics, however, gives no grounds for counterposing the former to the latter.[4]

Thus Soviet writings define the systems view in such a way as to put it on a legitimate and different level from Marxism-Leninism.

The emphasis on systems analysis in Soviet management theory, not to mention its more formal applications in the military and economic sphere, was to be followed by a series of works on systems analysis of global development and the systems "approach" to international relations research.[5] The application of systems thinking to international relations followed its application to domestic politics. In 1970 Fedor Burlatsky noted that "politics also began to be examined as a system and was elaborated as a kind of control centre of society."[6]

Only in the 70s did the "science of international relations" become a legitimate independent discipline in the Soviet curriculum. Largely due to the holistic character of Marxist ideology, a lengthy debate normally precedes the formal declaration of any focus of study in the Soviet curriculum as an "independent discipline." Since the early 60s the Soviet leadership has made considerable efforts to institutionalize the foreign policy-making process, increasing the participation of specialists in the political decision-making process. There are now seven institutes involved in research on international relations within the framework of the Academy system alone. Currently the Soviet Union has approximately three times as many academic specialists working on U.S. foreign policy as Americans working on Soviet foreign policy.[7]

The institutes under the Soviet Academy of Sciences, however, apparently are not conducting much research on systems modeling of international relations. During the 70s there was a "Model Building Laboratory" at the Institute of USA and Canada of the Soviet Academy of Sciences; but it was disbanded.

The two most active centers of systems research of global development and international relations in the Soviet Union are the All-Union Institute of Systems Research (VNIISI) and the Center for Systems Analysis of International Relations at the Moscow Institute of International Relations (MGIMO). The former is under the joint supervision of the Soviet Academy of Sciences and the State Committee on Science and Technology (GKNT). The latter is under the education and training section of the Ministry of Foreign Affairs.

Several studies have been published recently in the Center for Systems Analysis of International Relations at the Moscow State Institute of International Relations (MGIMO). Researchers at the Center have engaged in mathematical modeling of political development.[8] There have also been some innovative models of the negotiations process, including a model on the thinking patterns of Otto Von Bismarck in the late nineteenth century, which is discussed below.[9]

A substantial Soviet contribution on the systems approach to international relations is contained in two publications by E. A. Pozdniakov. Not surprisingly, Pozdniakov gives all credit to Marx for defining systems concepts: "Marx gave the classic example of the analysis of a socio-economic formation as a complex system, possessing its structure, its special laws of function and transition to a new qualitative condition."[10] Pozdniakov employs the terms of contemporary systems thinking, even acknowledging the value of Western terminology such as homeostasis and equilibrium:

> Definite work on the elaboration of new methodological premises, and on making existing premises more precise, has been conducted by systems specialists in various areas of knowledge (both abroad and in our country). One can include the principles of integrality, structurality, homeostasis (the ability of a system to support and preserve a stable internal equilibrium (*ravnovesie*) despite a "disturbing action" of environmental factors), the interaction between a system and its elements, the system and the environment, etc.[11]

Yet Pozdniakov hastens to disparage the conservative worldview of Western systems thinking:

> In essence, the structure is the result of a definite mode of system activity. Applied, let us say, to the system of interstate relations the revelation of structure and law-governed regularities of its function allow cognition of the mechanism of the vital activity of the system, namely: the conditions of the preservation and support of dynamic equilibrium, the conditions of stability of the system, the mechanism of structural crises, and the correlation of forces in the system.
>
> It behooves one to underline, however, that the structural-functional approach is oriented mainly to the study of those changes in the system which are not tied with its radical transformation, with fundamental qualitative changes inside the structure. It researches the mechanism of vital activity of the system taken in a certain time "slice", not in its development, not in the process of the transition of the system from one of its qualitative conditions to another.
>
> The Marxist method of the research of complex systems is not limited by the study only of structural-functional relations and ties. Its most important principle in the research of social phenomena is the principle of historicism...[12]

Thus Pozdniakov asserts that researchers must begin with class analysis; moreover, Western systems theorists are said to be in a methodological deadlock (*tupik*). Emphasizing the "two-plane" characterization of the systems approach and Marxism-Leninism, Pozdniakov writes in the closing line of a 1976 journal article: "...for Marxist scholars, armed with the singularly scientific dialectical method, the systems approach is one of the most important means of the cognition of social phenomena (including even those as complex as international relations)."[13]

Soviet research on global modeling has been centered in the All-Union Institute of Systems Research (VNIISI) since its creation in 1976 within the Academy system. VNIISI's long-time director, Academician Dzhermen Gvishiani, was appointed in late 1985 to the position of Deputy Chairman of Gosplan. Since 1977 VNIISI has published a journal of collected works on "systems modeling of global development processes."[14] VNIISI does not, however, focus exclusively on global modeling. Researchers at VNIISI are involved in various projects, including those involving artificial intelligence.[15] Recently VNIISI has produced some exploratory systems analyses of international relations.

The VNIISI publications characterize international conflict as a function of international socio-economic contradictions and avoid language suggesting homeostasis or Bukharin's Marxian theory of equilibrium—a theory which effectively reduces contradiction to conflict. Gvishiani stresses that the goals of system management are different in socialist and capitalist countries due to class differences; at the same time he asserts that the historical dialectic is transforming the world into an "integral social system."

Along with this basic affirmation of Marxist historicism, Gvishiani puts unusual stress on "the growth of the subjective factor," the need for conscious human action and guidance to change the world. Almost restating Marx's 11th thesis on Feuerbach, Gvishiani notes that "humanity cannot limit itself to simple observation and explanation of [the dangerous consequences] of the scientific-technological process," it must engage in global modeling with application of formal mathematical methods to evaluate possible alternate courses of development and correctly manage this development.[16] Soviet writers contrast the active stance of socialist science with an alleged Western stress on fatalism, the "unmanageability" (*neupravliaemost'*) of scientific-technological development. Though in general Soviets comment favorably on the Forrester/Meadows world models, sometimes echoing the "Limits to Growth" theme,[17] they note that "the fundamental inadequancy, however, of [the Forrester and Meadows models] is the highly limited conception of the possibilities of the conscious influence of humanity on its own development."[18] Of course, in this regard the Soviets protest the lack of historical optimism embodied in the *Limits to Growth* study and suggest alternatively a belief that technological innovation can increase the efficiency of the production function beyond what Meadows, et al., assumed. The basic faith in the power of technology to assist humanity in progressing "forward" pervades the Soviet literature on global modeling.

As far as content, the VNIISI studies of the international system emphasize general themes such as reducing inequalities between North and South and ending the arms race.[19] One recent VNIISI paper includes a chart of various scenarios of the development of international relations.[20] The scenarios can develop along three lines: mutually beneficial cooperation, confrontation between the developed and developing nations, or a division of the developing nations into blocs around the developed countries. The outcomes are a function of three main variables: 1) the path of development of the developing countries (revolutionary-democratic or capitalist); 2) the tendency of world development (detente or confrontation); and 3) the relations among the developed capitalist nations (whether they follow a policy independent from the U.S. or whether the U.S. dominates). The "optimal" scenario, according to the Soviet author, involves a structural reconstruction of the international division of labor by means of greater integration of developing countries into the world economy; to facilitate this process it is necessary that the majority of developing countries choose a "revolutionary democratic path."

2 COMPARISON OF U.S. AND SOVIET APPROACHES TO MODELING INTERNATIONAL CONFLICT

In principle, there are two basic ways to approach the phenomenon of international conflict. One approach would focus on the internal sources of conflict (material need for resources, economic need for markets, demographic need to occupy additional territory, etc.). An opposite approach would focus on the conflict behavior generated by the interaction of one state with another (psychological insecurity and consequent arms build-up by one state induced by the growing arsenal of another state, desire of an aggrieved state for revenge against another, desire of one state to dominate another).

Given the highly divergent political cultures of the U.S. and Soviet Union, it is not surprising that in general scholars in the two countries tend to approach the analysis of international conflict from divergent angles. Soviet analysts stress the determinative role of the socio-economic structure of a nation, and adhere to the Marxian tenet that contradiction is a fundamental characteristic of capitalist society. Consequently, Soviet analysts predictably focus on the internal determinants of conflict in Western states, downplaying the conflict-creating factors that arise when states interact. American analysts, on the other hand, do not share the Marxist historicist worldview, and tend to stress the subjective aspects of international conflict, focusing on conflict generated in the process of interaction of states through arms races, misperception, misunderstanding, and other psychological factors. Ultimately, however, these two approaches can be seen as complementary, as will be developed below.

That the generalizations about the different U.S. and Soviet approaches to international conflict are basically valid is evidenced by the following analysis which identifies two major points where the approaches typically diverge. First of all, Soviets object to Western "general" conflict models. Secondly, in contrast to many Western analyses, Soviet writings traditionally accord little significance to psychological factors—though this has changed dramatically in recent years.

2.1 The Possibility of a General Conflict Model

If you assume that states with different socio-economic structures are not likely to behave in fundamentally different ways, even in historically different situations, then you can construct a general theory of international conflict. The classic American analysis of international conflict, the Richardson arms race model, is a general conflict model. Richardson writes:

> This theory is about general tendencies common to all nations; about how they resent defiance, how they suspect defense to be concealed aggression, how they respond to imports by sending out exports; about how expenditure on armaments is restrained by the difficulty of paying for them; and, lastly, about grievances and their queer irrational ways, so that a halting apology may be received as though it were an added insult.[21]

The Richardson model includes a variable for "grievances" which, according to the author, "could include deeply rooted prejudices, standing grievances, old unsatisfied ambitions, wicked and persistent dreams of world conquest, or, on the contrary, a permanent feeling of contentment."[22] The Richardson arms race model also incorporates the motive of fear, which moves nations to increase their armaments because of the existence of the arms of opposing nations. The Richardson arms race model does not in any way define hostility as a function of a state's social system, nor does it explicitly anticipate socio-economic structural evolution over time.

Soviet analyses, on the other hand, focus primarily on the dynamic socio-economic structure of states in a given conflict. Soviet specialists typically decry models such as the Richardson model as ahistorical, suggesting that

> ...the efforts of the bourgeois political scientists to create a "general theory of conflict" is connected in the final analysis with the sufficiently transparent intention of bourgeois science to elaborate a new "general" theory of development, in which they would clearly like to substitute the conception of contradiction with the category of conflict...[23]

Soviet publications must emphasize the basic Marxian concept of contradiction as the source of all antagonism; contradiction inheres in specific social systems or arises among states of different social systems.

The central nature of the concept of contradiction is evidenced by the standard Soviet typology of international conflict. All Soviet conflict typologies are time-specific, i.e., they are said to be valid for a particular historical epoch. Soviet analyses currently divide the world into three subsystems: the socialist, capitalist and developing world.[24] The Soviet typology of international conflict includes the following three basic types:

1. international conflicts "directly reflecting contradictions between the two social systems in the world arena;"
2. international conflicts "arising from the efforts of imperialism to suppress peoples' liberation movements;"
3. international conflicts "reflecting inter-imperialist contradictions and struggle."[25]

The Sino-Soviet conflict raised a particularly uncomfortable problem for Soviet theoreticians: according to Marxist-Leninist theory, there are no "objective contradictions" in relations among socialist states. The Soviets typically treat the Sino-Soviet conflict as anomolous, refraining from defining a new type of "inter-socialist" conflict.

Focusing on the motive force which allegedly arises from Marxian social contradictions between the types noted above, Soviet analyses typically belittle the value of Richardson-type arms race models. A Soviet critique of a book by a Finnish scholar makes this point clear:

> ...Vayrynen tries to base his research on the conception of "action-reaction", interpreting the arms race as an automatic reaction of one party to the military preparation of the other, raising the phenomenon of "action-reaction" to the rank of a fatal factor. The Finnish scholar does not, however, fully support the above-noted conception and as a result of his work admits that its scientific value for the understanding of the process of militarization is "highly relative". The author's use of the method of factor analysis turns out to be more fruitful, allowing him to overcome the abstractness of traditional bourgeois treatments of the arms race and the reasons for international conflict, to uncover the special purposes of the military-industrial complexes of capitalist states. Using significant factual material, R. Vayrynen shows the fundamental lines of the fusion of monopoly with the state apparatus in capitalist countries, analyzing the economic base of the military-industrial complexes, examining the means of its influence on the process of political decision-making.[26]

Thus Soviet writings consider as "scientific" only those studies which focus on the "economic base of the military-industrial complexes" in *capitalist* countries. Given the alleged militaristic nature of capitalist states, there arises a "contradiction" with socialist states. To accept as central the psychological action-reaction explanation of international conflict, Soviets would have to reduce the Marxian notion of "contradiction" as the source of all conflict to a notion of conflict as the disturbance of equilibrium among states.

The above Soviet critique further reveals—perhaps unwittingly—that the Richardson arms race model is not without utility. The Soviet Union, being a socialist state, by definition cannot be the source of conflict; but it does respond to external threats in a fashion not unlike the pattern elucidated by the Richardson arms race model. The Soviet critique asserts that

> ...failure to consider the class nature of militarization, his reduction of it exclusively to purely economic categories (the absolute level of military expenditures of states, the amount of national resources devoted to military purposes, etc.) makes more difficult an adequate reflection of the real state of affairs...the author, in particular, presents the defensive policy of the U.S.S.R. in a distorted perspective, although he admits that the actions of the Soviet Union in the area of armaments was more an "answer" than an "initiative."[27]

Soviet writings typically accord legitimacy to the Richardson-type arms race explanation only with regard to the response of the Soviet Union to allegedly aggressive actions by Western states. Some Soviet writings on the history of the Russian state specifically identify the action-reaction phenomenon as the primary reason for Russian military action:

> ...the primary, constant factor that influenced the formation of the Russian people's national character, the Russian state system and Russian socio-political thought was the constant military pressure to which Russia was subjected as a result of being sandwiched between Europe and Asia and the Russian people's corresponding armed resistance to that pressure...[28]

2.2 The Role of Social-Psychological Factors

According to Marxian ontology, conflict lies in objective reality, not in people's heads.[29] Thus soviet writings have traditionally disparaged the role of social-psychological factors. Yet, as will be developed below, Soviets have in recent years accorded more significance to the subjective element in international conflict, noting that social-psychological elements may become "relatively independent" and determinative in certain conflict situations.

A typical Soviet analysis in the early 70s noted that the choice of tactical devices in a conflict

> ...is dictated first of all by economic, political and class factors, and not merely by a certain "strategic account," by the psychology of those who run countries—the participants of a conflict...[30]

To focus on psychological factors, from a Marxian point of view, is to emphasize secondary, dependent variables—"epiphenomena" of the historical process.

Yet recent Soviet writings increasingly give weight to social-psychological elements. The most recent Soviet study of international conflict notes that "in the final analysis all factors act through mediation of the social-psychological dimension."[31] Thus, according to Marxism-Leninism, human consciousness reflects objective reality, albeit with significant distortions induced by individual, group and class consciousness. Within this theoretical framework, Soviets can then legitimately discuss social-psychological factors in conflict. Moreover, in acute international crises, Soviet writings now say that the social-psychological element may take on a life of its own:

> In the period of acute international crises, many foreign policy decisions, as the experience of many capitalist countries has shown, are made in a more narrow circle of statesmen and may carry the imprint of hasty, emotional and spontaneous thinking, based on insufficient or unverified information, on certain individual peculiarities of these statesmen, at times of a psychological character. All this introduces a special subjectivism.[32]

Soviet writing now stresses that once a crisis has emerged on the basis of "objective" contradictions, it acquires its own "relatively independent" logic which gives it traits of "uncontrollability" and "unmanageability." The recent Soviet study on international conflict devotes an entire section to the role of social-psychological factors. The section discusses the socio-psychological determinants of the decision-making of the leadership (always of capitalist or developing countries—never of a socialist country). It describes the psychological conditions in which conflict escalation occurs:

- the impossibility for the leadership to achieve the goals it has set before itself by a nonconflictual path;
- the impossibility, for objective or subjective internal reasons, to change the character or priority of goals before the political leadership;
- the actual or seeming superiority of "its" state in the area of capacity to conduct a given conflict.[33]

One group of Soviet researchers at the Center for Systems Analysis of International Relations have recently developed a model of the thinking patterns of Otto Von

Bismarck.[34] The study considers each statesman as a carrier of specific class, group and personal demands. These demands are determined by content analysis of historical documents. The model itself simulates the process by which each statesman endeavors to realize his goals. The model simulates the element of lower or higher stakes in a crisis by varying the effect a given input will have on different statesmen depending on their particular foreign policy aims.

2.3 Synthesizing the Two Approaches

The above analysis suggests that U.S. and Soviet specialists tend to approach the problem of modeling international conflict from divergent angles. Yet, as was already noted, the approaches can be seen as complementary and ultimately convergent.

While the Soviet approach begins with the conflict-generating mechanism within [capitalist] states, it has increasingly incorporated social-psychological factors that arise in the process of interaction among states. This has occurred along with the general trend in Soviet writings on the "systems approach" to international relations to give increasing weight to themes of interdependence and the "relatively independent" effect of the international system on nation-states.[35]

Regarding U.S. approaches, there are many American models that do not fall into the general type epitomized by the Richardson arms race model. A prominent example of a basically nonreactive Western model is the Choucri-North "lateral pressure" model.[36] The lateral pressure model focuses on the way in which structural variables in society determine propensities for external expansion. The model is non-Marxian, focusing on the interactive effects of population growth and advances in technology, the development of military and other capabilities that affect national power. The lateral pressure model also incorporates the action-reaction phenomenon: domestic pressures may evoke a violent response by rival powers. One recent American model suggests that the arms race formulation is empirically deficient in accounting for the spending patterns of the major nations. It suggests that domestic politics play a major role in spending patterns—as the Soviets would assert regarding the U.S. The study, however, suggests that for *both* the U.S. and the U.S.S.R. the military allocation process is used to influence domestic political and economic conditions.[37]

Thus both American and Soviet models have come to incorporate insights gained from seeing international conflict as a function of internal as well as interactive factors. A currently existing European model, the Luterbacher SIMPEST model, perhaps best incorporates the elements characteristic of U.S. and Soviet approaches. The model incorporates national peculiarities, type of economy and political regime, as well as interactive factors.[38]

Regarding general approach to modeling, Soviet analysts seem to share many of the same premises as Western scholars, at least in the realm of global modeling. Articles in VNIISI publications note that a model may not—indeed cannot—encompass all influential elements and yet may have significant heuristic value if not full predictive value. In this regard, the contemporary Soviet view diverges widely from the child-like optimism of its earliest international systems thinker, Bukharin, who in his day expected that science would be able to predict dates of future social revolution in light of technological development.

ENDNOTES

1. For an overview, see Schwartz, S., "Recent Soviet Adaptations of Systems Theory to Administration Theory," *Journal of Comparative Administration*, Vol. 5, August 1973; Beissenger, M., *The Politics of Convergence: the diffusion of Western Management Ideas in the Soviet Union*, doctoral dissertation, Harvard University, September 1982.
2. See Tuzhilin, A., *Systems Analysis and Mathematical Modeling in the U.S.S.R.* Monograph Series on the Soviet Union, Delphic Associates, November 1983, p. vi.
3. See for example: Gvishiani, D., "Marksizm-leninizm i global'nye problemy" *Modelirovanie protsessov global' nogo razvitiia: sbornik trudov* Issue 8, VNIISI, Moscow, 1979, p. 12.
4. Shakhnazarov, G., "Vzgliad v budushchee" (Moscow, 1978) English translation: "A Look into the Future" (Moscow: Novosti, 1978), p. 39.
5. Soviets distinguish between the systems "approach" and systems "analysis" by defining the former as involving general principles and the latter involving the formal method of elaborating specific complex problems in the spheres of management, military strategy and weapons systems and the resolution of concrete socio-economic problems.
6. Burlatsky, F. M., *Lenin, gosudarstva, politika* (Moscow, 1970).
7. Legvold, R., "The Study of Soviet Foreign Policy: The State of the Field and the Role of IREX" in "Foreign Area Research in the National Interest: American and Soviet Perspectives," IREX Occasional Paper, Vol. 1, no. 8, 1982, p. 27.
8. Kozhemiakov, A. C., Sergeev, V. M., "Prognozirovanie politicheskogo razvitiia: osobennosti razrabotki dinamicheskikh modelei" *Sotsiologicheskie issledovaniia* no. 1, 1981, p. 112.
9. Lukov, V. B., and Sergeev, V. M., "Opyt modelirovaniia myshleniia istoricheskikh deiatelei: Otto Fon Bismark, 1866-1876" *Voprosy Kibernetiki: logika rassuzhdenii i ee modelirovanie*, Scientific Council on the Complex Problem "Cybernetics", Academy of Sciences, (Moscow: 1983), pp. 149-161.
10. Pozdniakov, E., "Sistemnyi podkhod v issledovaniiakh mezhdunarodnikh otnoshenii" *Mirovaia ekonomika i mezhunarodnye otnosheiia*, no. 11, November 1975, p. 82.
11. *Ibid.*, p. 83.
12. *Ibid.*, p. 84.
13. *Ibid.*, p. 91.
14. See Gvishiani, D. M., "Metodologicheskie problemy modelirovaniia global'nogo razvitiia". Preprint, VNIISI, Moscow, 1977; "Metodologicheskie problemy modelirovaniia global'nogo razvitiia" *Voprosy filosofii* no. 2, 1978; "Global'noe modelirovanie: kompleksnyi analiz mirovogo razvitiia" *Problemy mira i sotsialisma*, no. 8, 1978; and "Marksizm-leninism i global'nye problemy" in the collection "Modelirovanie global'nogo razvitiia" 8th issue, pp. 5-17, VNIISI, Moscow, 1978.
15. One VNIISI project focuses on the creation of expert intelligence systems (SPEIS). (See Gelovani, V. A., Kovrigin, O. V., Smolianinov, N. D., "System for Producing Expert Intelligence Systems—SPEIS" Institute for Systems Studies, unpublished paper, Moscow).
16. Gvishiani, D., "Metodologicheskie problemy modelirovaniia global'nogo razvitiia" *Voprosy filosofii*, no. 2, 1970, p. 13.
17. Cf. Dubovskii, S. V., "Chelovek-mashinnyi issledovaniia global'nikh problem" *Chelovek-mashinnaia sistema modelirovaniia protsessov global' nogo razvitiia: sbornik trudov* VNIISI, Moscow, 1982, p. 5.
18. *Op cit*., Gvishiani, D. M., "Metodologicheskie problemy...", p. 16.
19. Cf. Katsura, A. V., Novik, I. B., "Mirovozzrencheskie osnovaniia cheloveko-machinnogo modelirovaniia" in *Chelovek-mashinnaiia sistemia modelirovaniia postsessov global' nogo razvitiia:sbornik trudov*, VNIISI, Moscow, 1982, p. 19. These themes are said to be objective necessities because the old world economic order no longer corresponds to the real correlation of forces and productive-technological structure of world economic ties. In order to meet the "objective demands" of the current age, the world economic order must be reconstructed in the direction of the liquidation of economic inequalities of the developing countries in relation to the industrially developed capitalist states.
20. Bogoroditsky, V. K., Pirogov, G. G., "Novyi mezhdunarodnyi ekonomicheskii poriadok i modelirovanie global'nogo razvitiia" Preprint, VNIISI, Moscow, 1982, p. 53.

21. Richardson, L. F., *Arms and Insecurity* (Pittsburg, 1960) p. 13.
22. *Ibid.*, p. 16.
23. Gantman, V., (ed.) *Mezhdunarodnye konflikty sovremennosti* (Moscow: izd. Nauka, 1983) p. 9-10.
24. Bogoroditsky, V. K. and Prigorov, G. G., "Novyi mezhdunarodnxi ekonomicheskii poriadok i modelirovanie global'nogo razvitiia" Preprint, Vsesoiuznyi nauchno-issledovatel'skii institut sistemnikh issledovanii (VNIISI) 1982, 62 pp.
25. *Op cit.*, Gantman, V. See section on typology of conflict (pp. 35-47).
26. Tiulin, I., "Teoriia konflikta i problema sokhraneniia mira" *Mirovaia ekonomika i mezhdunarodnye otnoshenia* No. 9, 1974, pp. 139-141. This includes a critique of Raimo Vayrynen, *Militarization, Conflict behavior and Interaction*, (Finland, 1973) and Ferry de Kerckhove, *Les Tactiques des conflicts* (Quebec, 1972).
27. *Ibid.*, p. 140.
28. Nesterov, F., *Sviaz' vremion*, (Moscow, 1980).
29. According to Lenin's definition in *Materialism and Empiriocriticism*, the Soviets define "objective" as that which exists outside and independent of human consciousness.
30. *op cit.*, Tiulin, p. 141.
31. Gantman (ed.) *op cit.*, p. 223.
32. *Ibid.*, p. 58.
33. *Ibid.*, p. 220.
34. Lukov, V. B., Sergeev, V. M., "Opyt modelirovaniia myshleniia istoricheskikh deiatelei: Otto Fon Bismark, 1866-1877" *Voprosy kibernetiki: logika rassuzhdenii i ee modelirovanie*, Moscow, 1983, pp. 148-161.
35. See, for example, Pozdniakov *Sistemnyi podkhod*, p. 39.
36. See Choucri, N., and North, R., *Nations in Conflict* (Cambridge, 1975).
37. Cusack, T. R., and Ward, M. D., "Military Spending in the United States, Soviet Union and the People's Republic of China," *Journal of Conflict Resolution*, Vol. 25, no. 3, 1981, pp. 429-469.
38. The government sector demands resources from the economic sector and uses them in accordance with conditions dictated by internal politics. The government sector makes decisions in questions of security, arms expenditures, foreign trade, etc. The actions of the government sector are directly linked to the actions of the government sectors of other countries.

SELECTED BIBLIOGRAPHY

Books and Monographs

Alymov, A., (ed.) *Sistemnyi analiz i modelirovanie razvitiia i razmeshcheniia proizvoditel'nikh sil*, (Kiev, 1975), 151 pp.

Blauberg, I., Yudin, E., *Stanovlenie i sushchnost' sistemnogo podkhoda* (Moscow, 1973).

Choucri, N., North, R., *Nations in Conflict* (Cambridge, 1975).

Gorskii, D., (ed.) *Sistemnyi analiz i nauchnoe znanie* (Moscow: Izd. Nauka, 1978), 245 pp.

Kulik, V., *Sistemnyi analiz* (Kiev, 1974), 107 pp.

Pozdniakov, E., *Sistemnyi podkhod i mezhdunarodnye otnosheniia* (Moscow: Izd. Nauka, 1976).

Richardson, L., *Arms and Insecurity* (Pittsburg, 1960).

Shakhnazarov, G., "Vzgliad v budushchee" (Moscow: Novosti, 1978), 63 pp.

Susiluoto, I., *The Origins and Development of Systems Thinking in the Soviet Union: political and philosophical controversies from Bogdanov and Bukharin to present-day re-evaluations* (Helsinki: Suomalainen Tiedeakatemia, 1983), 196 pp.

Tuzhilin, A., *Systems Analysis and Mathematical Modeling in the U.S.S.R.*, Monograph Series on the Soviet Union, Delphi Associates, November 1983, 133 pp.

Articles

Bogoroditskii, B., Pirogov, G., "Novyi mezhdunarodnyi ekonomicheskii poriadok i modelirovanie global'nogo razvitiia" Preprint, VNIISI, Moscow, 1982, 63 pp.

Cusack and, T., Ward, M., "Military Spending in the United States, Soviet Union and the People's Republic of China: *Journal of Conflict Resolution* Vol. 25, no. 3, 1981, pp. 429-469.

Ermolenko, D., "O sotsiologicheskikh issledovaniiakh mezhdunarodnikh otnoshenii" *Sotsiologicheskie issledovania*, no. 2, 1983, pp. 37-43.

Gelovani, V., Kovrigin, O., Smolianinov, N., "System for Producing Expert Intelligence Systems—SPEIS" VNIISI, unpublished paper, Moscow.

Goriachev, A., Sergeev, V., "Peregovory kak zadacha diskretnogo programmirovaniia" *Technicheskaiia kibernetika*, No. 6, 1980, pp. 22-26.

Graham, L., "Cybernetics" in *Science and Philosophy in the Soviet Union* (London, 1973), pp. 324-354.

Gvishiani, D., "Marksizm-Leninizm i global'nye problemy" *Modelirovanie protsessov global'nogo razvitiia: sbornik trudov*, Issue 8, VNIISI, Moscow, 1979, pp. 5-17.

Gvishiani, D., "Metodologicheskie problemy modelirovaniia global'nogo razvitii" *Voprosy filosofii*, no. 2, 1978, pp. 14-28.

Kedrov, B., "Printsip istorisizma i ego prilozhenii k sistemnomu analizu razvitiia nauka" *Sistemnye issledovaniia. Ezhegodnik.* Institut istorii estestvoznaniia i tekhnika, Moscow, 1973, pp. 5-18.

Kozhemiakov, A., Sergeev, V., "Prognozirovaniie politicheskogo razvitiia: osobennosti razrabotki dinamicheskikh modelei" *Sotsiologicheskie issledovaniia* No. 1, 1981, pp. 111-115.

Leibin, V., "Mirovozzrenchskie osnovaniia global'nogo modelirovaniia" *Sistemnoe modelirovanie protsessov global'nogo razvitiia: sbornik trudov*, Issue 14, Moscow, 1980, pp. 4-18.

Lukov, V., Sergeev, V., "Opyt modelirovaniia myshleniia istoricheskikh deiatelei: Otto Fon Bismarck, 1866-1876" *Voprosy kibernetiki: logika rassuzhdenii i ee modelirovanie* Moscow, 1983, pp. 148-161.

Novik, I., "Gnoseologicheskie aspekty global'nogo modelirovaniia" *Modelirovanie protsessov global'nogo razvitiia: sbornik trudov*, Issue 8, VNIISI, Moscow, 1979, pp. 17-28.

Pozdniakov, E., "Sistemnyi podkhod v issledovaniiakh mezhdunarodnikh otnoshenii" *Mirovaia ekonomika i mezhdunarodnye otnoshenia*, no. 11, November 1975, pp. 82-91.

Schwartz, D., "Recent Soviet Adaptations of Systems Theory to Administrative Theory" *Journal of Comparative Administration*, Vol. 5, August 1973, pp. 233-264.

Takhtadzhian "Tektologiia: istoriia i problemy" *Sistemnye Issledovaniia* Issue 2, 1971, pp. 200-278.

Timofeev, I., "Obsuzhdenie ezhegodnika *Sistemnye issledovaniia*" *Voprosy filosofii* no. 3, 1977, pp. 154-61.

Tiulin, I., "Teoriia konflikta i problema sokhraneniia mira" *Mirovaia ekonomika i mezhdunarodnye otnosheniia*, no. 9, 1974, pp. 139-141.

DISCUSSION OF ALLYN'S PRESENTATION

Von Glasersfeld

Only a little time is left for solving the problem of our survival. Assume that we manage to go on for another forty years *without* nuclear war. In 2025 there will be more than just the threat of starvation to the entire globe, unless we do something now about population growth. In 2025 all the "civilized" nations will have poisoned their environment far beyond the point of possible return, unless we do something now.

Both of these problems threaten the U.S. as well as the USSR. Neither of these problems can be solved without immediate and intensive collaboration. Yet, they are not even considered in the top-level meetings (e.g., Geneva). If discussions and joint action were seriously initiated, the absurdity of nuclear war would very soon be understood by all concerned.

Mitroff

I couldn't say it better.

Umpleby

I think that this discussion may be off the mark. I believe that anyone who has thought about nuclear war for even a moment knows that if it were to occur it would be an unimaginable disaster. Yet, knowing this, we go on building nuclear weapons and threatening their use. Why? I suspect the answer may be that at present we have no means of resolving serious disputes between the superpowers except to threaten cataclysmic destruction. I think a useful question is, what do we need to know about how people in the U.S. and the USSR approach issues of conflict and compromise in order to be able to create an alternative way of resolving disputes?

Ben-Eli

I sense that our concerns are shifting from methods to purposes.

Lapin

Of course, any model has limited capacities, but it must be decided how to use these capacities. It is especially important to develop a general concept of the world and scenarios for its development. We thought it important to introduce the "global system" concept which will help in considering the interaction of man and nature. This system has its history and is changing. Today it is somewhat more developed and not a simple additive system, but it is not an organic unity. Now it is a complex of systems consisting of contradictory but interconnected components, whose interdependence is increasing.

The contradictions in the global system can be seen as a set of problems. Three levels can be distinguished.

Most General Alternatives

Global Problems <----------> Universal Problems

↓ ↑

Sectoral Problems <----------> Regional Problems

"Universal problems" are ones which exist in all or the majority of countries but can be resolved by these countries' own efforts (e.g., illiteracy). "Global problems" require joint efforts by many or most countries (prevention of nuclear war, epidemic diseases).

We had a group of experts analyze the interconnection among 22 problems and they isolated 3 complexes of problems: general systems problems, economic problems, and social problems.

The interconnections of the problems sharply impedes the solution of any one of them in isolation. The co-evolution variant seems promising; but the problem of preserving peace in the world is the main one.

Rapoport

There are rules of subordination among countries. There are rules of security. Models make it easier to reach agreement. But are we modeling behavior or relationships?

Allyn's models are very interesting and useful for ordering our thinking, but we have to remember that any model does not reflect reality adequately. This is why we must develop our methodological tools. I propose to separate the behavior of social systems and the relationships between them as subject matters for our investigations. The latter defines the former and are more important. Relationships are more stable and are based on "the rules of the game." They are understandable for politicians and managers. We have good opportunities for modeling relationships and their structures using as a basis the systems approach.

Lapin

I support Professor Rapoport's opinion of the importance of studying relations and not simply behavior. This gives rise to the need to model profound underlying historical processes which have lasted for hundreds and thousands of years. These processes are cyclical, and knowledge of these long-term trends will enable us to make more reliable predictions for the next twenty to thirty years.

Seven Scenarios for the Future is the title of a study that was done in the USSR at the Institute for Systems Studies.

CHAPTER

15

THE DEVELOPMENT OF ERGONOMICS AS A SYSTEM DISCIPLINE

Vladimir M. Munipov

For scientific disciplines which are still in the making, and where the object of science is to a significant extent still being formulated, the search for general principles to guide the study of related activities is foremost. Ergonomics (human factors engineering) is this kind of discipline. Therefore, discussing the problem of formulating the knowledge system for this discipline reveals certain specifics. Ergonomics as a science is a complex study of the interaction of a man (or a group of people) with hardware and software components, the object of the activity and the environment in the process of attaining the purpose of the activity or while preparing to perform it. The object of ergonomics consists in raising the efficiency of man's activities and the functioning of "man-machine" systems while preserving the health and satisfaction of a man.

The task of ergonomics as a sphere of practical activity consists in designing and improving the processes (ways, algorithms, methods) of performing activities and the methods of preparation (education, training, adaptation) for it, including characteristics of means and conditions of activities immediately affecting the efficiency of the activity and the psycho-physiological condition of a man.

The comparison of ergonomics and classical scientific disciplines allows us to come to the conclusion that there is a profound difference between them. Moreover, judging by classic examples of scientific knowledge, which were formed by the end of the 19th century (for instance mathematics and the natural sciences), ergonomics cannot be considered a scientific discipline. M. de Montmollin, defining ergonomics as technology, proceeded from similar assumptions (Montmollin, 1967).

A question, which was formulated by D. Meister, is also extremely significant. It is reasonable to ask whether an orientation to the laboratory which was developed in the 19th century to imitate the physical sciences is the most effective framework for behavioral research (Meister, 1982).

At the same time ergonomics possesses many attributes of a scientific discipline, as a specific sphere of scientific activities, distinguished from other ones. One can refer, for example, to the existence of particular organizations and groups of professional researchers, specializing only in the field of ergonomics. Most important is the fact that ergonomics is actively influencing the practice of research and the design of man-machine systems.

Ergonomics is not developed according to the pattern of "a research line—a field of research—a scientific discipline," which is characteristic of the classical sciences, originating from the relevant basic scientific disciplines. Ergonomics is rather developed according to the pattern of "a scientific movement—a complex scientific discipline."

The development of ergonomic research has its pre-history, which could be reconstructed. On the one hand, we could identify the component parts of the present-day ergonomics,—these are A. A. Bogdanov's "tectology," A. K. Gastev's concept of "labour aims" (The History of the Soviet Psychology of Labour, 1983), T. Kotarbinsky's "praxeology," V. M. Bekhterev's and V. N. Miasishev's ideas on "ergonology" (The History of the Soviet Psychology of Labour, 1983), and present theoretical premises for a number of fundamental ergonomic postulations. On the other hand, ergonomic knowledge includes many particular principles and guidelines, which have been developed by applied, theoretical, and experimental research on work activities (Taylor System, engineering psychology, and physiology of labour). However, one cannot conclude that ergonomics evolved merely by uniting these lines and concepts.

Ergonomic postulations are not limited by the use of notions and methods of other scientific disciplines aimed at investigating a man and his work, such as sociology, psychology, physiology, hygiene, anthropology, etc., as well as man-machine systems, cybernetics, systems engineering, systems analysis, etc. The following question arises: What kind of new "addition" turns this "conglomeration" of information from various disciplines into a specific knowledge—proper ergonomic knowledge? This "addition" seems to be a way of arranging knowledge. That is, the knowledge from other disciplines, related to certain scientific and technological tasks, makes up a new integral whole.

The implication is not a mere sum of various knowledge. It always has a choice character and involves some changing of the borrowed notions. That is why the methods of ergonomics are not homogeneous in principle. They cannot be related to any single scientific discipline, taken separately.

If we consider the requirements of the systems approach, in particular those which are essential for building systems of knowledge (E. G. Yudin, 1978), then we should pay attention to a feature such as the hierarchical arrangement of the knowledge system. This is ensured not by a simple description of the relations between elements of knowledge, but by singling out some system generating connections among them. Identification of such a "hierarchy," in order to establish a system of ergonomic knowledge, is possible if one interprets an activity category as the main system generating principle of the organization of this field of science.

Sometimes an apprehension is expressed that founding ergonomics on one principle will lead to an "impoverishment" of the discipline. However, such an apprehension is hardly grounded. Accepting a certain principle as a system-generating one for a particular discipline is by no means identical to limiting this science to a single category. The question is not to reduce the contents (of ergonomics in this case) to the analysis of the activity, but rather to build, with the help of this category, an explanation of all of the phenomena under investigation and to find a logic in their relationships and interdependence.

The activity in ergonomics is not only an object of scientific research but also becomes an object of control. It may be arranged in a harmonious system of functioning and/or developed on the basis of a set of fixed principles which should be formulated in ergonomics, social psychology and labour sociology. The activity in ergonomics is also an object of design, that is, people in the field have the task of identifying ways and conditions of optimum implementation of work and professional training activities (preferably new ones). Finally, ergonomics activity becomes an object of multilateral estimation, which should be made according to various criteria, such as efficiency, reliability, work satisfaction, comfort, etc. Thus, ergonomics activity becomes the beginning, the middle and the end of ergonomic analysis, organization, design and estimation (Zintchenko, Munipov, 1979).

To solve theoretical and practical tasks of ergonomics the activity notion should be given a certain constructive meaning. But the task is not as simple as that, since ergonomics up to now has had no conceptual schemes of activity analysis of its own. Therefore, we widely use in ergonomics some conceptual schemes of activity analysis from related disciplines, from psychology and sociology in particular. These conceptual schemes are not only used, but are also transformed by ergonomics according to the specifics of the tasks being worked on. Ergonomics confronts the necessity of developing methods of analysis and of identifying the functional structures of various activities: from relatively elementary to the very complex ones which have emerged as a result of the scientific and technological revolution. This is an obligatory prerequisite for activity optimization, for rationally designing its new forms. Otherwise, these tasks are solved either on the basis of common sense or by trying a great number of factors, which in some way influence efficiency and other activity aspects, that is, by the method of successive approximation.

The development of the activity category by L. S. Vygotsky, S. L. Rubinstein, A. N. Leontiev (Vygotsky, 1960; Rubinstein, 1940; Leontiev, 1975; The concept of activity in Soviet psychology, 1979) and other Soviet learned men marked the beginning of psychology in the Soviet Union. The development of the activity category in psychology also had an important influence on the formation of N. A. Bernstein's physiological concept of constructing movements. He tried to build a bridge from the "physiology of reactions, which for a long time monopolized the attention of psychophysiologists, to the physiology of activity" (Bernstein, 1966).

Using psychological ideas allows us to make an essential step forward in founding ergonomics. Researchers in a number of countries are working in this direction. D. Meister believes that further development of human factors research will depend, to a great extent, on work on methodological problems, many of which have much in common with similar problems in present day American psychology (Meister, 1981). He stresses that human factors researchers should understand that without a serious scientific foundation for their methods, their practical work becomes an art but not a science. A list of the most important problems of human factors research, formulated by D. Meister, shows that the most important ones are directly or indirectly concerned with human behavior in the system being designed (Meister, 1982). The development of a new line of research in ergonomic psychology is made possible by the increasing significance of human activity research in ergonomics, which "is the beginning and the end of ergonomic research, estimation and design," as is correctly stated by Soviet learned men, according to J. Leplat (Leplat, 1980).

The great heuristic value of the psychological theory of activities is proven by the fact that the activity idea provided explanations of various social phenomena and contributed to the development of a number of sciences and fields of technological practice, including sociology, linguistics, the history of science and industrial design.

The idea of "activity" can have two meanings relative to ergonomics. On the one hand, human activities are treated as its object. On the other hand, the means of investigation are, by their nature, activities themselves. The first level of ideas we can call the object order. This is a description of various types of man-machine systems (MMS), ideas regarding the character and regularities of human activities within MMS, etc. These notions reflect the reality under investigation. They are generalized analogues (or models) of the reality which is studied by ergonomics.

Regarding the arrangement of ergonomic knowledge, one should note that it is impossible to fully characterize the totality of ergonomic propositions by means of only one level, by the notions of the object order. They make up only one part of ergonomics, although a very essential one. Another important part of ergonomics includes the various models, methods, etc., which describe how to act with ergonomic objects (i.e. how to analyze, study and design them).

We could define this second level of ergonomic notions, using an analogy with the first level, as notions regarding research, thus pointing out their relation to another reality, not to the reality under investigation but to the methods for working with it. By way of example, one can mention such notions as "relationship analysis," "error classification," "listing of functions," etc.

Thus, ergonomic knowledge consists of two levels describing two types of reality. It requires simultaneously knowledge of the object and knowledge of how to study and design this object, the two types of knowledge being organically interconnected (Alekseyev, 1977).

It should be noted that these two levels of knowledge are characteristic not only of ergonomics but also of a number of scientific disciplines as well, if they are oriented to the analysis of complex objects. For example, the systems analysis of complex objects (in solving management problems in particular) has at its center the notion of scenario, which implies a specific type of model. The scenario does not just present a related field of real objects which should be considered by the potential management decision. According to R. Ackoff, the scenario is "an artificial reality" or "a scheme of the method." It reflects, in some sense, alternative strategies of management decision making (indicated by diagrams, block-schemes, etc.).

This idea of the two levels of ergonomic knowledge allows us to imagine the following type of structure: the first object-level reflects a reality under investigation, but not the whole of it, and only a closed one, the character of the problem being determined by the second level of ergonomic notions. In other words, ergonomic knowledge is arranged in every particular case by the task of research, i.e. by the ideas regarding the potential methods of solution.

Having as its main object of research a man-machine system, ergonomics studies its particular properties, known as human factors, which present the integral characteristics of the man and machine relationship. These characteristics manifest themselves under certain conditions of system functioning aimed at particular objectives. The above definition shows that characteristics and properties covered by the human factors notion are a quality of the system, not some isolated attribute of man-machine system components. As compared to the latter, human factors are properties of the second order, arising as a result of integrating into one entity those natural properties which characterize the environment, the objective properties of the machine (technological means) and functional properties of the human being, including social ones.

Ergonomics is not interested in all possible "primary" properties of man, machine and environment, but only in those which are defined by the position and the role of man in a system. That is why they are called human factors. This does not means that they are few. Evidently, those models will be most perfect and efficient in which the number of man's properties, functionally related to his position in a system and with its natural and objective properties, would reach the maximum.

Ergonomics should ensure the definition of a necessary and sufficient nomenclature of functional relations between man-machine system components, since only in this case can a system acquire the status of a system, having a present efficiency and meeting certain requirements and criteria. The nomenclature of functional relations should be constructive (but not endless) and should meet a number of criteria for man-machine system estimation, both technologically (stability, reliability, noise immunity), socially and economically. It specifically expresses itself in the fact that ergonomics does not simply operate with various sets of primary properties, qualities and attributes (hygienic, physiological, psychological, social-psychological, technological, ecological, etc.) taken from other disciplines. Ergonomics also transforms them as system characteristics by establishing required functional relations among them.

Human factors, treated as the most important characteristics of a man-machine system, thus become a superposition of primary properties and some fixed (or dynamic) functional relations among elements and components of a man-machine system.

Human factors are not homogeneous. To single out and classify them is a complex task. It is important to note that human factors by themselves act as structural formations at various levels of complexity.

Thus, it is man's activities that are the basis for singling out human factors which should be taken into consideration while defining functional relations between man-machine system components. It is also true, that such functional relations are a necessary basis for the organization of successful human activities within a system.

The activity category becomes a theoretical foundation of the above understanding of human factors. Since it has been developed as a systems discipline, ergonomics contributes to further development of the systems approach and the activities approach. "As a consequence of the failure to study actual systems," says D. Meister, "there are serious deficiences in behavioral knowledge. At the most fundamental level, for example, we lack a system taxonomy that would permit us to classify various types of systems in which they occur" (Meister, 1981).

It is impossible for ergonomics to separate an object of research ("man-machine systems"), which is a "performance object," from the influence of the research activities, of designing and improving this object. At the same time the two types of activities are continuously interacting and complementing each other without fusing into one.

Since ergonomics has to do with a "performance object" of research and design, it becomes necessary to make compatible systems and performance notions; and this results in the necessity of making compatible generalized structural, functional and performance schemes by way of a single whole description. This, in its turn, stipulates specifics of the ergonomic design of the above objects. And here the simulation method is an important one. Ergonomics essentially influences a progressive development of the general systems problems and results in enriching the methodological tools of the systems approach.

REFERENCES

1. Alekseyev, N. G., The Two-ply character of ergonomic knowledge and the systems approach, *Activity and psychic processes*, Moscow, 1977, pp. 81-87.
2. Bernstein, N. A., *Essays on physiology of movements and physiology of activity*, Meditsina Publishers, Moscow, 1966.
3. Wersch, J. V. and M. S. Sharpe (eds.), *The concept of activity in Soviet psychology*, Armonk, New York, 1979.
4 Zintchenko, V. P., V. M. Munipov, and O. G. Noskova (eds.), *The history of the Soviet psychology of labour*. Papers (the 20-30-es of the XX century), Moscow University Publishers, Moscow, 1983.

5. Leontiev, A. N., *Activity, Consciousness, Personality*, Politizdat Publishers, Moscow, 1975.
6. Leplat, J., *La psychologie ergonomique*, Paris, 1980.
7. Meister, D., *Behavioral research and government policy: Civilian and military*, Pergamon Press, New York, 1981.
8. Meister, D., The role of human factors in system development, *Applied ergonomics*, 1982, vol. 13, N 3.
9. Montmollin, M., *Les systems hommes-machines*, Paris, 1967.
10. Rubinstein, S. L., *The problems of general psychology*, Moscow, 1940.
11. Vygotsky, L. S., *Development of higher psychic functions*, Prosveshcheniye Publishers, Moscow, 1960.
12. Yudin, E. G., *System approach and principle of activity*, Moscow, 1978.
13. Zintchenko, V. P., Munipov, V. M., *The principles of ergonomics*, Moscow University Publishers, Moscow, 1979.

DISCUSSION OF MUNIPOV'S PRESENTATION

Umpleby

If I understand you correctly, you are emphasizing the connections between ergonomics and systems theory.

Munipov

Yes. Ergonomics is similar to systems research in that it is interdisciplinary and functions, at least in part, at a different level of analysis than traditional scientific disciplines. Because ergonomics is a relatively new field, we tend to reflect more on the nature of knowledge in our field than perhaps do people in more well-established fields.

CHAPTER

16

CYBERNETIC TOOLS FOR MANAGEMENT: THEIR USEFULNESS AND LIMITATIONS

Michael U. Ben-Eli

1 GENERAL SYSTEMS THEORY, CYBERNETICS, AND MANAGEMENT

When Stuart Umpleby called to invite me to participate in this meeting, he suggested that a useful contribution would involve a discussion of the application of the systems approach and Management Cybernetics in my consulting experience. Of course, to do so in the allotted time of 20 minutes will force me to commit the cardinal sin of systems thinking, that is, to be a ruthless reductionist. I thus stand the unfortunate chance of doing a great disservice not only to the principles and methods, but to the actual cases themselves.

Nevertheless, I shall try to simplify without completely trivializing and use a number of management related cases to highlight a few ideas concerning approaches that are rooted in systems thinking and cybernetics. As I go along, I shall comment about their usefulness, effectiveness, and limitations. Although the context of my comments pertains specifically to management, many underlying ideas have been touched upon during the last two days in one way or another by speakers concerned with other seemingly unrelated topics. This fact in itself should give us some cause for satisfaction, validating, as it does, the unifying potential of the systems approach.

First, by way of orientation: I see systems thinking and cybernetics as two inextricably related bodies of thought that, not withstanding their precise history (concerning which, Professor Zeleny earlier shed some intriguing light), have burgeoned in the West since the forties and early fifties, largely due to the work of Ludwig von Bertalanffy and Norbert Wiener. These gentlemen, together with their associates and followers, laid the foundations for a new conceptual orientation and an intellectual tradition, many implications of which have still to find their way into guiding the manner in which we manage organizations and govern national and international affairs.

I choose to make the distinction between the systems approach and cybernetics in order to highlight the emphasis on general systems characteristics on the one hand and the more specifically focused attention on mechanisms of regulation on the other. Thus, if I were to distill the essence of each as pertinent to our discussion I would evoke some of the following ideas.

In relation to General Systems Theory there is the emphasis on wholeness and the relation of wholes and parts; the idea of connectivity; nestability; interdependence; synergy; and irreducibility. All of these point to fundamental characteristics of systems in general and highlight the idea that in dealing with systems strategies quite different from those effective in dealing with clock-like mechanisms are essential.

Cybernetics on the other hand focuses on the relationship between the structure of systems and their behavior, thus bringing to the fore the issues of structure, process, and organization. It also formulates general laws of regulation focusing on underlying mechanisms and, most recently, has concerned itself with recursive processes where the implications of self-observation and self-reference have to be taken into account.

The literature in both fields is vast and there is little purpose in dwelling on any particular aspect of it here. The point, however, is this: Over the last few decades a considerable body of thought related to the "system sciences" has evolved. It now ranges in scope from the philosophical and general to the methodological and specific. Taken as a whole it offers many useful tools for tackling problems encountered in Management situations. Note, incidentally, that I use the term Management with a capital M meaning the governance and proper guidance of human affairs—from the personal to the managerial and societal.

Here, if I may make use, or, since it was intended in a different context, possibly a misuse, of the framework suggested earlier by Professor Sadovsky, I would characterize the range of this chest of tools as follows:

- On the Philosophical level we have that particular conceptual framework usually referred to as systems thinking or the systems view of the world.
- On the General Scientific level we have the formulation of general laws, for example, in cybernetics, the laws of control, as exemplified by Ashby's Law of Requisite Variety.
- On the Specialized Scientific level we have, for example, the well developed mathematics of servo-mechanisms.

- Finally, on the level of "Methodics," there are various system-related techniques such as System Dynamics simulations, various methods of operations research, and the like.

This, of course, is only a crude way of saying that much is available on different levels. Yet, there are important limitations. Thus, for example, while the "philosophical" and "general scientific" levels offer particularly powerful guiding principles that help orient our thinking when dealing with complex systems, the more specific and well developed techniques are sometimes useful, but often not potent enough. Many non-trivial management situations involving high complexity, a high rate of change, the need for integrating many conflicting forces, and, in particular, the need for genuine innovation, creativity, and novel design, still defy most of the available analytical tools. The problem is not in the least one of lacking the appropriate epistemology. We still attempt to tackle complexity by reducing it to simple manipulable components. Therein, I suspect, also lies the cause for the emergence of baffling paradoxes of the kind referred to earlier by Mitroff.

Ultimately, the consultant, was well as the manager, depends on an approach that combines an art with a science. Assisted by available techniques, both have to find their way by relying on "feelings," intuition, and experience. I suspect, however, that managerial instincts could be sharpened significantly by effective exposure to systems thinking.

2 SELECTED CASES OF MANAGEMENT INTERVENTIONS

In describing our own experience at The Cybertec Consulting Group, with projects in which systems thinking and management cybernetics were particularly useful, I chose three cases, all carried out in the field of health care. One involves the assessment of the impacts of a proposed new program on the operations of a teaching hospital. The second involves the design and implementation of a comprehensive strategic planning effort at a leading academic medical center, and the third has to do with a management organization study at that same medical center and a number of other similar institutions. Let me touch briefly upon each.

a. The first case involved a major teaching hospital in New York City. Management in that institution developed some innovative program ideas intended to contribute to the overall quality of patient care, as well as to a cost reduction in hospitalization. The key feature of the new program related to the construction of a new sub-acute care facility to which patients would be transferred near the end of their stay, prior to discharge. The main argument was that given an average length of stay of ten days in the acute hospital, a transfer of convalescing patients for the last two or three days of their stay to a non-acute facility, where medical technology and nursing care would not be as intense, would reduce the overall cost of hospitalization.

At the same time, patients would benefit by staying in a more relaxed, less institutionalized environment where, together with a friend or a family member, a "care-partner," they could receive special education about their condition. Overall, the program would ease the transition from hospital to normal routine and ensure better post-hospital care at home.

The cost reduction argument appeared logical enough since, at the time, the reimbursement system in New York paid hospitals a set amount per patient day regardless of the intensity of service required. Thus, over an average of ten days, seven will be spent in the acute, costly facility and three in a less acute, less costly environment, thus reducing the average overall cost for the hospital for the given level of reimbursement.

This straightforward, linear view ignored the underlying systemic dynamics of the situation, which could potentially produce unintended results, some with serious financial implications. In fact, a patient transferred to the new sub-acute facility would result in an empty bed at the main hospital. Because this particular institution had a significant queue of patients waiting for admission, an empty bed would be immediately filled with a new admission representing a type of patient that normally requires considerable resources. The rate of patient throughput would accelerate and, as a result, the new program could actually translate into a considerable intensification in utilization of services. Thus, while average cost per episode would decline, the overall cost of operations could expand significantly. Since reimbursement levels remain unchanged, the resulting impact on the "bottom line" could be intolerable.

This was clearly a systemic problem because the actual level of intensification of services would largely depend on patterns of admissions and patient mix, and these, in turn, were likely to change with the introduction of the new sub-acute facility. Under different circumstances, different shifts in admission patterns and patient mix might result, giving rise to different levels of service intensification with related cost implications.

It became apparent that different management policies and patient protocols could greatly affect the outcome. Therefore, a better understanding of potential impacts, as well as the underlying dynamics of the various forces at play, were important if the clinical and economic viability of the whole program were to be assured. Further study was, therefore, encouraged and the vehicle used was a System Dynamics simulation model that made possible the study of impacts given various scenarios of patient profiles and management policies.

An extensive analysis of the critical factors affecting patient flows and occupancy balance between the two facilities was undertaken, revealing the intricate interactions among the key variables. Figures 1 and 2 highlight the underlying structure and dynamics of the system involved.

The model as a whole was conceived as a series of linked levels, shown in Figure 3. The first represented the underlying dynamics of critical factors affecting the flow between the two facilities. The second translated sets of simulated conditions into utilization patterns as manifest in operating room time, intensive care beds, diagnostic procedures, nursing services, and the like. The third computed the impact of changing utilization patterns on resource requirements and the last produced the resulting financial picture.

All in all, the development and use of the model enhanced management's understanding of the direct and indirect impacts of the new program. It was critical to the redefinition and refinement of some of the original concepts, it greatly facilitated

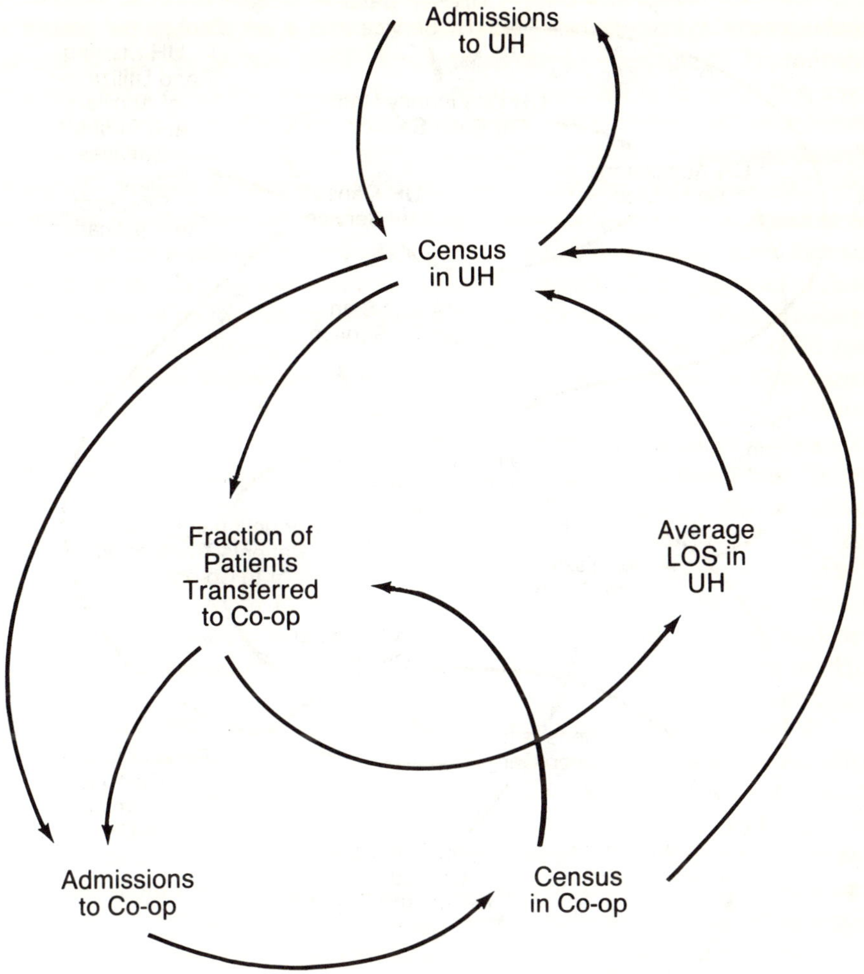

Figure 1 Several feedback loops affecting the census in University Hospital (UH) and the co-operative care facility.

implementation planning, and helped educate various groups in the hospital about the new program and its implications. The process of developing the model, undertaken closely with management, was particularly important in itself, since it forced key participants to clarify their thinking, allowed them to test continuously their assumptions, and produced an ongoing forum for dealing with differences of opinion and evolving an institutional consensus. All of these were as significant, perhaps ultimately even more so, as the actual numerical results of the simulations. This, in summary, was a clear case of a management problem where a systems approach backed by a well developed systems technique contributed greatly to understanding and guiding action.

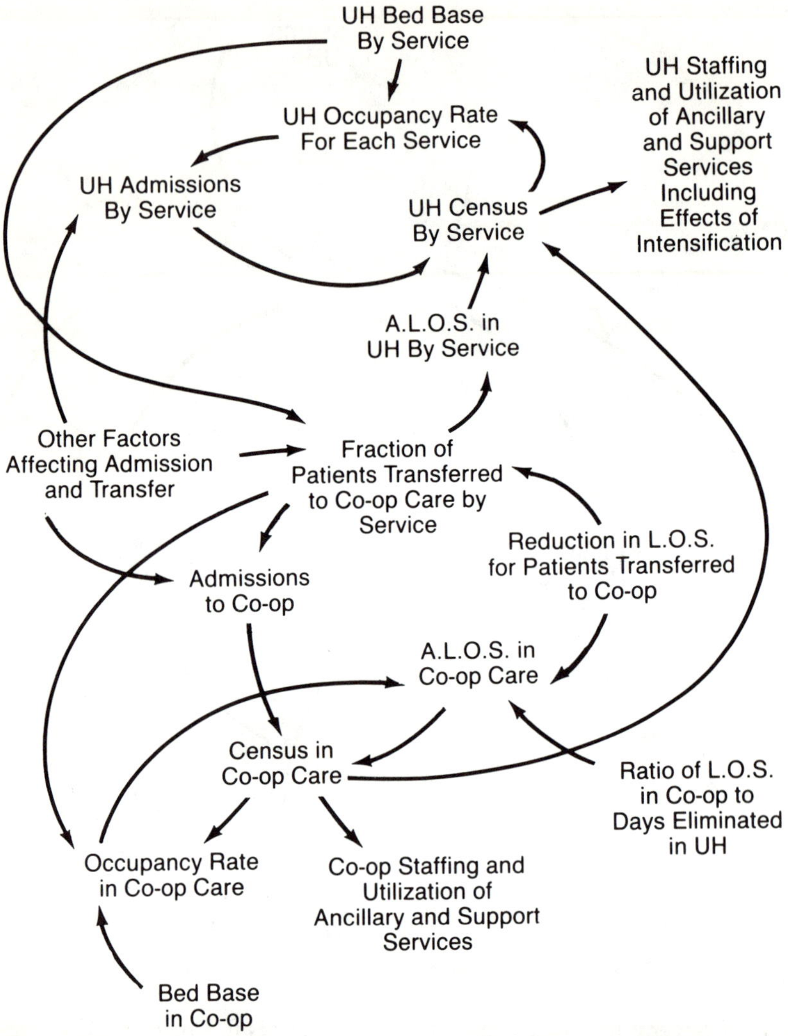

Figure 2 Examples of factors affecting occupancy balance and intensification of services.

b. The second case involved the design and subsequent implementation of a strategic planning process at a large academic medical center, also in New York. This particular institution consisted of a 1,200-bed hospital where some 36,000 in-patients where discharged each year, and some 250,000 clinic visits and 60,000 emergency visits were recorded annually. The center operated a medical school in which some 460 medical students were enrolled, it offered both Ph.D. and M.D./Ph.D. programs and was managing, on the average, some 900 funded research projects at any given time. The management of this complex institution, having just completed a three-year

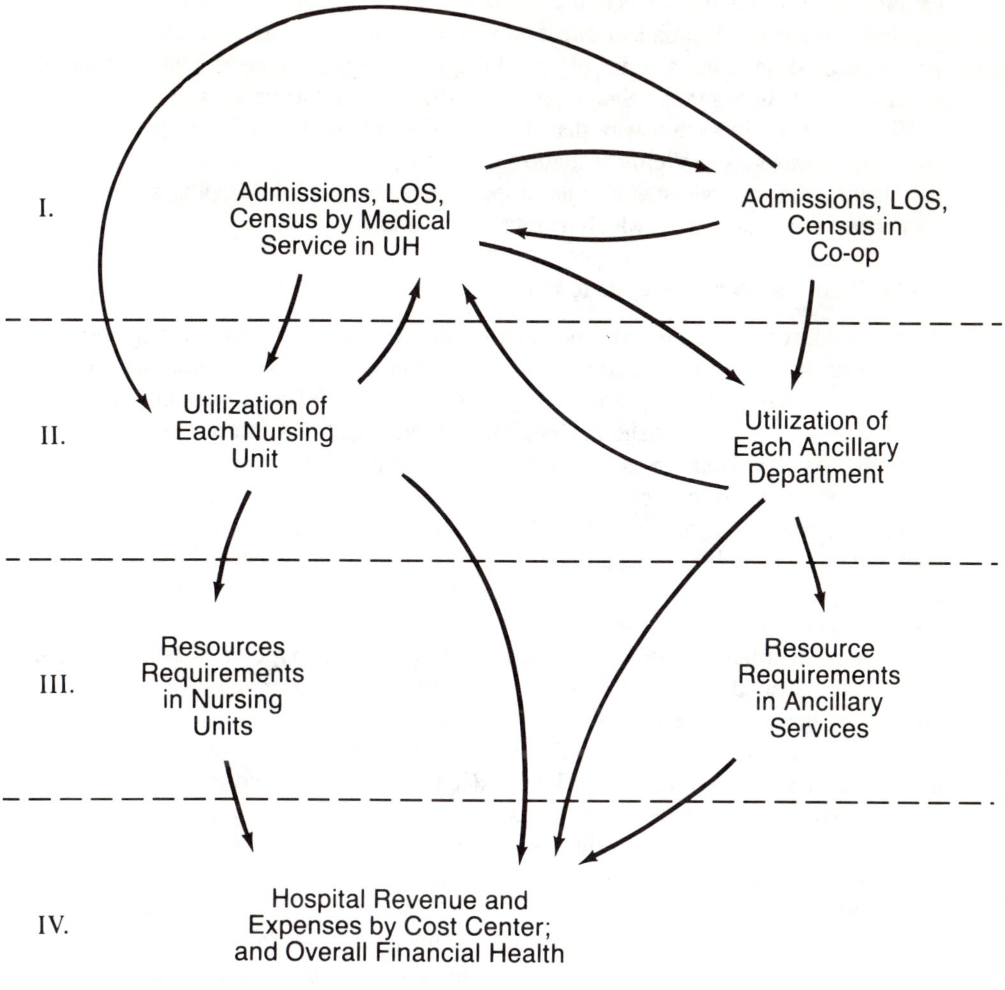

Figure 3 Model overview.

effort of stabilizing and shoring up operations, was ready to orient its views to the future. We were retained to assist them in developing their long-range plans.

The following concepts emerged as important guiding principles in designing and carrying out this strategic planning assignment.

- **A Whole System View of the Institution**

Early in the planning process, the tendency of various key players was to focus narrowly on isolated problem areas. Different groups of participants would emphasize this or that specific problem as being critical to the future and thus the most important to resolve. An important consideration in laying out the planning process, was, therefore, to shift the emphasis at the outset to a whole-system view of the institution, highlighting the key factors that affect its well-being and the way these interact. This was done by developing a series of visual diagrams, "system images," an example of which is shown in Figure 4. Single planning issues could thus be viewed as parts of a totality presented in such a way that the overall complexity and interdependency of key components was explicitly recognized. This was extremely useful in scoping out the planning effort, understanding its proper boundaries, and developing a consensus concerning the critical areas which ought to come under review.

- **The Planning Process was Issue Driven**

A complex institution such as the one involved is an extremely active "living" organization with its own tradition, internal culture, many, sometimes conflicting, constituencies, competing interests, and "a way of doing things." The momentum of its ongoing processes cannot be halted to await the development of a "master plan," nor can all institutional problem areas be dealt with at one time. Accordingly, the planning process was issue driven, meaning that very early attention was focused on identifying the most critical issues requiring long-term resolution, thus defining the targets for the planning process itself. By separating the critically important from the secondary and even imaginary problems, the early identification of key planning issues saved valuable time as the process unfolded.

Issue identification, incidently, derived from the usual type of external environment analysis on the one hand and an exhaustive analysis of various institutional profiles on the other. Perhaps the most important step at this phase, however, was dealing with the issue of purpose—the institutional mission. This was resisted at first as being a trivial pursuit, but was acknowledged to be of vital importance as soon as it was realized that fundamental differences of opinion existed among key players about policy issues concerning the institutional mission. Resolving these differences facilitated decision making later on when issues of program priorities, resource allocation, and the like had to be resolved.

As part of a competent planning process, the issue of institutional purpose, is, in my view, probably the most important. Around it revolves the whole question of institutional identity, its internal cohesion, and the very reason for its being. While this statement may appear to be too obvious to be of any consequence, the fact is that too often planning processes are data driven. The focus is on "information gathering" and the forest is easily lost for the trees. Data is useful for testing, validating, calibrating, and refining, but the underlying purpose provides the framework that gives it meaning. Purpose, after all, is a central driving force in social systems that by their very nature are "ideal-seeking." Confronting the issue of purpose may, at times, be painful, but dealing with it effectively can ultimately contribute significantly to the strength of the institution as a whole.

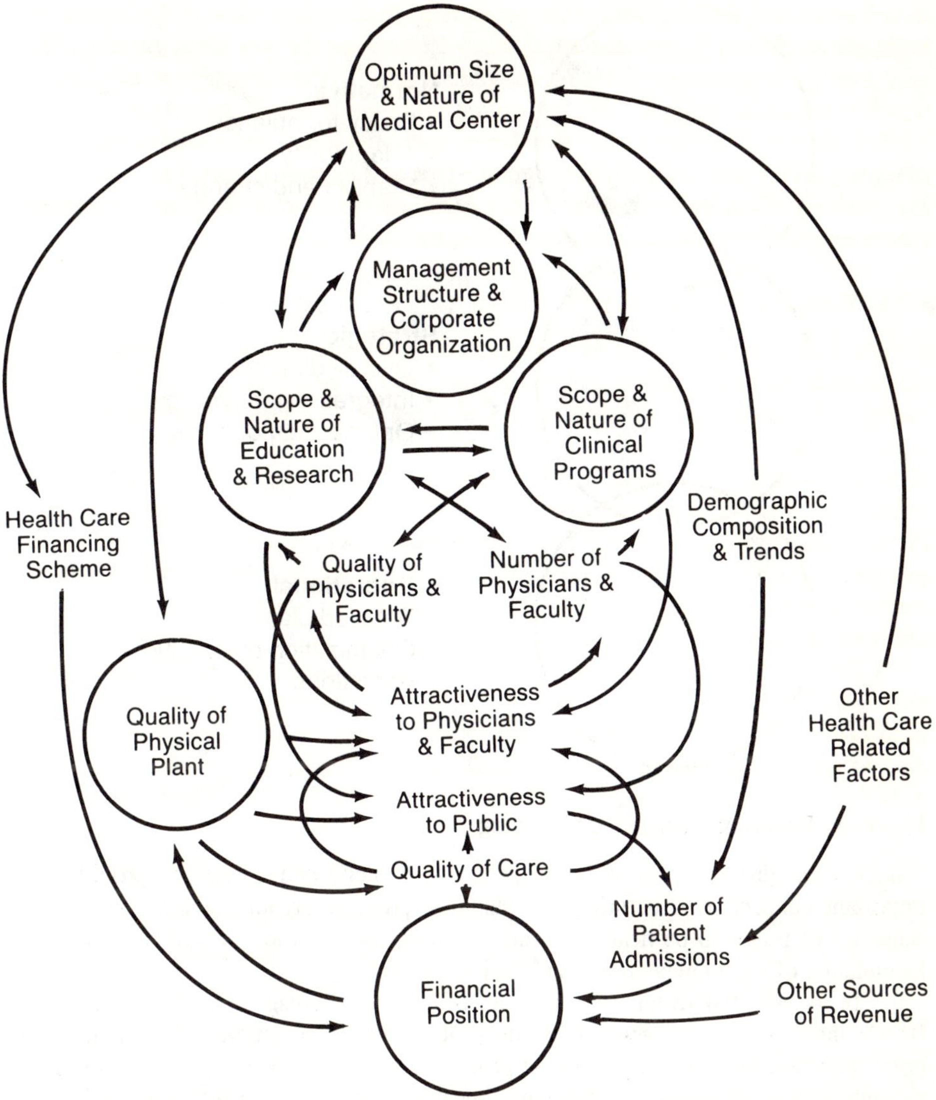

Figure 4 A system view of the medical center showing the interaction of some key variables.

• The Concept of Levels of Management

In meshing the planning process with the existing fabric of management the concept of levels of management, differentiating between logically and operationally distinct domains of management concern, offered a useful guiding principle. It is graphically depicted in Figures 5 and 6. The distinction between such levels of management

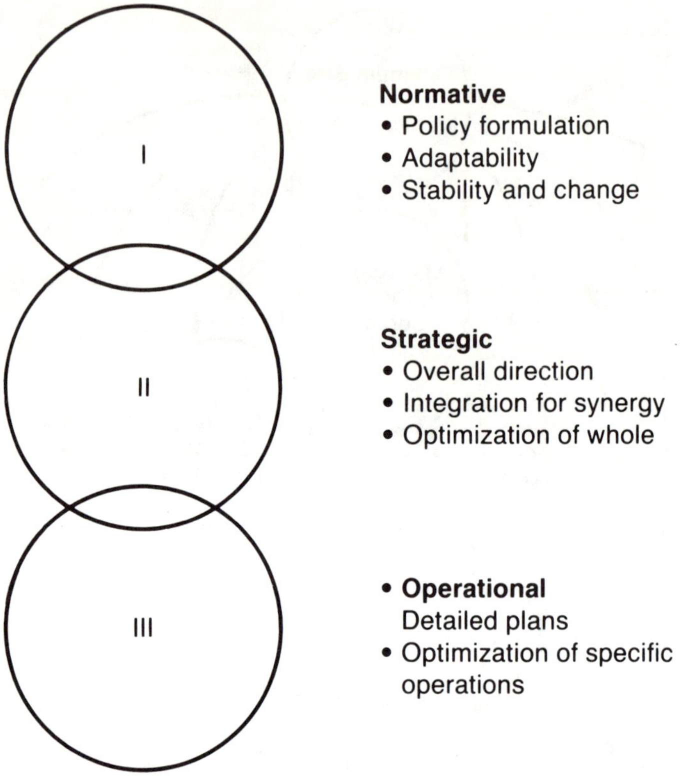

Figure 5 Levels of management/planning.

concerns, ranging from mission and policy to strategy and operations provided an important concept in structuring the planning process; organizing the content and sequence of issues to be dealt with; and defining the key fora for study, discussion, formulation of recommendations, and decision making.

In its essential features the concept of levels of management follows Stafford Beer's insights into the general anatomy of viable organizations. The underlying point, pertinent to structuring planning processes, is that approaching problems related to each such level requires a different conceptual orientation, a different method of handling, a different emphasis, a consideration of information at different levels of detail, and, more often than not, a different group within the organization.

With this view in mind planning issues can be organized as a two dimensional matrix: By functional type, such as financial, organizational, product related, and the like, as well as with respect to the proper level of management to which they logically belong—normative, strategic, or operational. Such a careful sorting out can be time consuming, but it serves well in helping to avoid the type of confusion that often results from inappropriate mixing of focus, logic, and language as they pertain to different levels of management concern.

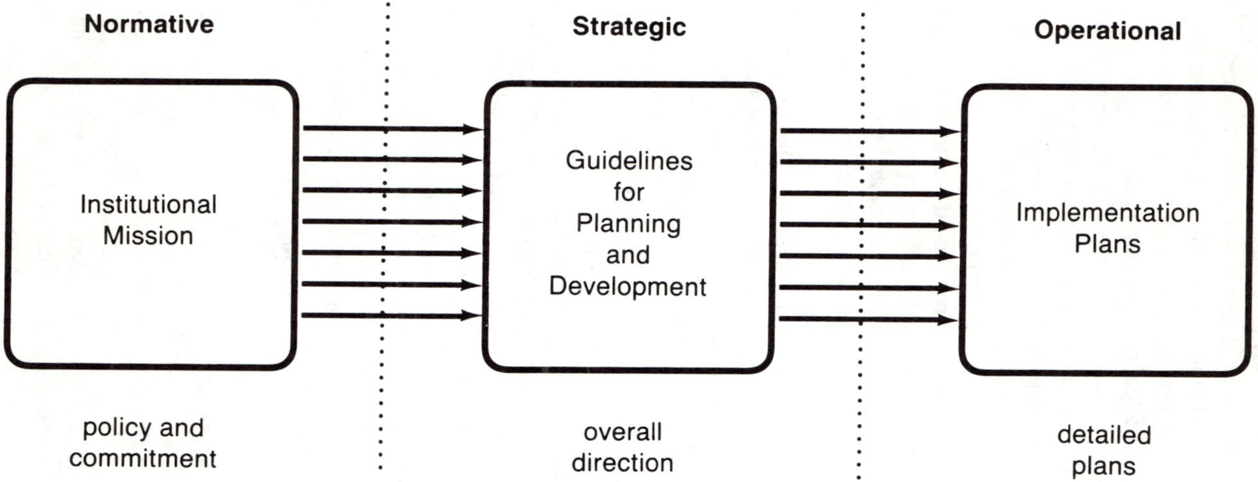

Figure 6 The planning process: conceptual sequence.

• **Participation and Underlying Structure**

Academic medical centers tend to be highly diversified pluralistic organizations where decision making and authority are diffused and where many key individuals play a number of different, sometimes conflicting, roles. This characteristic only reinforced our conviction that in order to be effective, the planning process should be organized to allow for a high degree of participation. An important element in the approach underlying this particular project was provided by the idea that key players—decision makers and those responsible for implementation alike—should be the actual authors of the plans, and that extensive open participation in formulation and refinement of ideas at all stages of the planning process were essential for building the consensus and the commitment required for successful implementation.

Some two hundred individuals ranging from board members and management to faculty, other important constituencies, and various specialty consultants were involved at one time or another. Orchestrating their work and harnessing their various contributions into a coherent effort was in itself an important task in the planning process.

Though popular in recent management literature, the idea of participative management or participative planning is not easy to apply. Beyond mere lip service it requires discipline, patience, and structure. To be effective, the energies such processes release must be carefully and thoughtfully channeled in ways that enhance creative results. From the view point of cybernetics, of course, broad based participation relates directly to the concept of making full use of an organization's potential variety and enhancing its own self-organizing capabilities.

The idea of underlying structure is also important. The structure is the vehicle through which the process actually takes place. Too often a well-functioning institutional structure for planning simply does not exist. It must be deliberately designed, put in place, and continuously reinforced. The functions of various groups involved need to be defined and the mode of their interaction specified. In this particular case a mechanism consistent with the concept of levels of management and reflecting the various functions and tasks required by the planning process was constructed. Essentially it consisted of a governance level steering committee, a senior management and faculty planning group, special institutional task forces, and a core group of planning staff and consultants providing the technical support and the overall management of the process itself. Overlapping membership was used to ensure continuity.

Properly designed and competently managed such a structure provides both for the necessary individual activities and for their continuous integration. It can act, in fact, as a giant homeostat mediating the various processes involved, facilitating the settlement of issues and the formation of institutional consensus.

In summary, the process as a whole proceeded in a sequence of four reiterative steps, moving from the general and open-ended to the ever more specific and committed. Planning issues were defined, their implications analyzed, a strategic direction was agreed upon, and was then given sharper focus in the formulation of a strategic plan. Implementation planning in all the essential areas of programs, facilities, finance, management, and organization followed. All in all, this project, unlike the previous one, did not rely on a specific system methodology alone. Here, a general systems orientation and the cybernetic view of management planning and control provided a guiding framework for the overall design and actual conduct.

c. The third case, which I can only touch upon very briefly, involved the redesign of an organization's management structure. This assignment was also carried out in the health care industry. I would only mention that this particular project relied heavily on using Beer's very powerful Viable System Model in diagnosing the weaknesses of a medical center's existing organization and synthesizing a design for a new one, more compatible with its future needs. For those of you who are familiar with Beer's work I would add that this particular project involved an overall system unification on a corporate level with some specific design interventions on levels III, IV, and V of the Viable System Model.

3 LIMITATIONS AND POSSIBILITIES

I hope that these brief examples are sufficiently convincing in buttressing my own conviction that general systems theory and cybernetics offer many important conceptual tools for dealing with management situations. But there are some difficulties. One in particular was alluded to earlier. It has to do with the notion that while some system techniques are too limited to be useful in complex management situations, others are too general to offer precise prescriptions for unambiguously resolving specific problems. The issue, it appears, is essentially an issue of requisite variety, namely, a poor match between existing models and the realities to which they pertain.

Two different attitudes are possible in this regard. One would assume that all we need are better, more refined theories, models, or equations, which will offer more powerful tools for prediction and control. With the aid of these, it can be argued, the difficulties and paradoxes encountered in dealing with complexity will eventually disappear. The other approach, more correct, in my view, would hold that there is an inherent uncertainty involved in genuine complexity that cannot be completely removed.

There are a number of reasons for such a claim. One has to do with the idea that management activities are exercised in the domain of social systems, a domain that is characterized by an enormous internal variety that negates the possibility of a single model or one single truth. Another reason relates to the relativistic aspect of the notion of "system." Systemic entities do not exist as an entirely objective part of an external reality. They are invoked, in fact, by observers and, therefore, their very definition is subject to myriad individual perceptual sensitivities, interpretations, motivations, and other "subjective" influences. Yet another reason pertains to the recursive characteristics of management processes whereby the observers practitioners are in fact elements of the system itself.

There are other perhaps more pragmatic and less esoteric difficulties that have manifested themselves in my own experience. One, for example, relates to the fact that in spite of half a century of developments in systems theory, the field is still somewhat peripheral to the mainstream of orthodox science. It certainly does not feature significantly in management education and its terms are considered too theoretical or too intellectual and are, therefore, regarded with some degree of suspicion. The system dynamics modelling effort described earlier, for example, was regarded from the outset as being somewhat eccentric, "experimental" at best, and although it ultimately helped shape important management decisions, it was funded as a special research activity and thus placed outside normal administrative operating costs.

In addition, there is the inevitable situation that much of the dynamics of management processes is dictated by considerations of power, internal politics, needs for self-preservation, and other motivations which, at times, have little to do with the problem at hand. While such forces may ultimately involve their own wisdom, they bear little sympathy to the rational problem solving orientation that is inherent to the systems approach. Last, but not least, is the somewhat sad observation that the mainstream of our management culture profoundly dislikes "problems"—the very term carries a negative connotation—perhaps since problem identification and subsequent resolution is likely to "rock the boat."

Notwithstanding these and possibly other difficulties, it is still my conviction that systems thinking and cybernetics can contribute even more than they already have to the understanding and management of human affairs. This would be done not necessarily by providing a calculus for predicting single events, but rather, by constructing qualitative "principles of explanations" (in the sense of von Foerster and Bateson), conceptual handles, with which one could come to grips with complexity that otherwise would elude one's understanding.

In the face of a world that is seemingly becoming ever more complex, many old approaches have been rendered useless. Evidence of mega-management failures are all around us. In the ever accelerating arms race, in chronic instability in the world economy and the monetary system, in the widening gap between the developed and the very poor countries, in the existing inequities of human rights, in the increasing destruction of vital parts of the planet's eco-system, and more. We could clearly benefit from a qualitatively new conceptual orientation, a new epistemology, and fresh new paradigms. Systems Theory and cybernetics could contribute greatly to their development.

In this context, it is important that we emphasize the need to learn to deal with complexity in an effective manner. We must become psychologically and culturally comfortable with the fact that complexity is a genuine, well nigh essential, part of our world. We must learn to relax insecurity-driven compulsions always to reduce complexity by imposing woefully inadequate low variety controls. We must learn to manage our enterprises, our society, our economy, and our planet as a whole, without trivializing them.

Ultimately, all this bears upon the chance of shaping a world where current and future generations can live safely, abundantly, and creatively, progressively increasing degrees of freedom and potential self-realization for all. In this regard, the underlying purpose, the vision, and the leap of faith, are often far more significant than any special case technique or analytical methodology.

Much has been said by other contributors about epistemology, about knowledge, and about the paradoxes which arise due to an inevitable incongruity between the world and our mental constructs of it. In the context of management and governance of societal affairs in particular, our conceptual/managerial tools are still far from adequate. For alas, or thankfully, perhaps, there is more to heaven and earth than is dreamt of in Horatio's philosophy.

I would like to close with a pertinent quote from Chapter Three of the Lankavatra Sutra, an old Buddhist text, much of which deals with epistemological issues. When asked by an aspiring Buddhisattva to talk about the nature of reality, the Buddha makes the following observation: "Things are not as they are seen, nor are they otherwise..."

REFERENCES

1. Ashby, R., *An Introduction to Cybernetics*, University Paperbacks, (1964).
2. Bateson, G., *Mind and Nature*, Dutton, (1979).
3. Beer, S., *Brain of the Firm*, Allen Lane, (1972).
4. Beer, S., *The Heart of Enterprise*, John Wiley & Sons, (1979).
5. Bertalanffy, L. von, *General Systems Theory*, George Braziller, (1969).
6. Forrester, J. W., *Principles of Systems*, Wright-Allen Press, (1968).
7. Suzuki, D. T., trans., *The Lankavatra Sutra*, Prajna Press, (1978).
8. Von Foerster, H., "Principles of Self-Organization in a Socio-Managerial Context," in H. Urlich and G. Probst (eds.), *Self-Organization and Management of Social Systems*, Springer Verlag, (1984).

DISCUSSION OF BEN-ELI'S PRESENTATION

Sadovsky

Sorry, but I do not agree with your interpretation of the scheme of methodological levels which I proposed in my presentation. From my point of view the systems approach, cybernetics, as well as the axiomatic method for example, belong to the level of general scientific methodology and not to the level of philosophical methodology. The reason for this is the following: Philosophical methodology has great meaning for one's world view, while the systems approach, etc. in general are irrelevant to this meaning. Analogously, the methods of reconstructability analysis created by George Klir are from my point of view a typical example of models which we can apply on the level of general scientific methodology. That is why I do not agree with considering them as belonging to the level of methods.

Ben-Eli

This was just a quick attempt to apply the levels to one field—systems and cybernetics—where you can make distinctions among contributions ranging from the philosophical through the "general" scientific and all the way to methods and technics. I agree that my interpretation may not correspond with your intention, and I will gladly withdraw the analogy, still pointing out though that contributions in our field range from the philosophical to the technical/methodological.

Mitroff

Philosophy leads the social and behavioral sciences, but philosophers are also led.

Sadovsky

The systems approach and philosophy have changed radically in recent decades.

Mitroff

Thomas Kuhn's work led to social-psychological approaches to philosophy.

Klir

Computers have also had an impact on philosophy.

Sadovsky

But philosophy remains the search for ultimate foundations and models.

Reckmeyer

We have moved from thinking of applications of theories to interrelationships between problems and philosophies.

Zeleny

It is difficult for theories to penetrate society. Economic challenges are easier to understand. The Japanese have the idea, "If it ain't broken, improve it." But the day of general systems theory is coming.

Mitroff

Even in the Midwest, managers are not asking for computer models.

Zeleny

Problems are now ahead of theories. There are no textbooks on how to manage in a robotized factory.

Mitroff

The Saturn Division of GM is a large scale experiment. There is a recognition that GM cannot build new ideas into cars within the old bureaucratic culture.

Rapoport

Management change is expensive. There is a need for stability. Similar approaches are being used in the Soviet Union and the U.S. It is important to create a scientific basis for this work. Of course there is also art in management.

Mitroff

The Japanese have learned how to adapt.

CONCLUDING REMARKS

Von Glasersfeld

Before the meeting closes, let me make a rather personal statement. These two and a half days have been immensely valuable to me, and I suspect that I am not alone in this feeling. There was not only talk but sincere interaction. All of us, I believe, have gained and will go away with our thinking expanded. Above all, we have made fruitful personal contacts that we want to maintain in the future.

I enthusiastically second Ian Mitroff's request that we should consolidate what we have achieved here and make sure that there will be many more meetings of this kind.

In that regard I would like to make a suggestion. Perhaps the most important single thing cybernetics has achieved since its birth is to have introduced into the scientific world the concepts of goal-directed behavior and purpose. I propose that we exploit this and make a commitment to discuss global goals at our next meeting. At this particular moment in time, it would seem of the utmost importance that, in spite of all political confrontation, we come to terms about global goals and decide how to preserve our planet. There is the most urgent need to get together and to agree on a joint effort to avert disasters that threaten us regardless of where we live and whatever nation we happen to belong to. The poisoning of the world's air and oceans, the destruction of vegetation and water supplies, and the growth of population that is already creating catastrophic regional starvation are threatening the survival of any civilization irrespective of political system or ideology. Unless we decide jointly to work against the exponential growth of these devastating processes, they will obliterate our future as surely as would the nuclear holocaust we are busy trying to avoid.

Mitroff

There are high level complex goals that unite us. What divides us can be discussed without rancor.

Reckmeyer

Means now exist for achieving global goals.

Sadovsky

Before this meeting only a few of us knew each other. Now we know each other much better. I hope the same people will be at the next meeting. We have two years to prepare. In the meantime we can have an intensive activity to prepare ourselves.

APPENDIX

NAMES AND ADDRESSES OF PARTICIPANTS

Bruce Allyn
Harvard Negotiation Project
500 Pound Hall
Harvard Law School
Cambridge, MA 02138

Igor Alekseyevich Apokin
Institute of History of Science and Technology
USSR Academy of Sciences
Staropansky Pereulok
1/5 103012 Moscow, USSR

Michael U. Ben-Eli
The Cybertec Consulting Group
1323 Madison Avenue
New York, NY 10028

George Klir
Department of Systems Science
Thomas J. Watson School
State University of New York
Binghamton, NY 13901

Oleg Alekseyevich Kossov
Institute for Systems Studies
9, Prospect 60 let Oktjabria
117312 Moscow, USSR

Nikolai Ivanovich Lapin
Institute of Philosophy
USSR Academy of Sciences
ul. of Volkhonka 14
121019 Moscow, USSR

Igor Andreyevich Lavrov
USSR Academy of Sciences
Leninsky Prospect 14
117901 Moscow, USSR

Ian I. Mitroff
Graduate School of Business
University of Southern California
University Park
Los Angeles, CA 90089

Vladmir Ivanovich Munipov
Institute for Industrial Design
State Committee for Science and Technology
129223 Moscow, USSR

Dmitry Aleksandrovich Pospelov
Computer Center of the USSR Academy of Sciences
117967 Moscow, USSR

Vladimir Samuilovich Rapoport
Institute for Systems Studies
9, Prospect 60 let Oktjabria
117312 Moscow, USSR

William J. Reckmeyer
Cybernetic Systems Program
San Jose State University
San Jose, CA 95192-0113

Vadim Nikolayevich Sadovsky
Institute for Systems Studies
9, Prospect 60 let Oktjabria
117312 Moscow, USSR

Stuart A. Umpleby
Department of Management Science
George Washington University
Washington, DC 20052

Ernst von Glasersfeld
Cognitive Processes Research
Department of Physics
University of Massachusetts
Amherst, MA 01002

Milan Zeleny
Fordham University at Lincoln Center
GBA-626 E
New York, NY 10023

INDEX

DB